EQUINE SAFETY

EQUINE SAFETY

STEPHEN A. MACKENZIE

THOMSON
DELMAR LEARNING

Australia Canada Mexico Singapore Spain United Kingdom United States

NOTICE TO THE READER

Publisher does not warrant or guarantee any of the products described herein or perform any independent analysis in connection with any of the product information contained herein. Publisher does not assume and expressly disclaims, any obligation to obtain and include information other than that provided to it by the manufacturer.

The reader is expressly warned to consider and adopt all safety precautions that might be indicated by the activities described herein and to avoid all potential hazards. By following the instructions contained herein, the reader willingly assumes all risks in connection with such instructions.

The publisher makes no representations or warranties of any kind, including, but not limited to, the warranties of fitness for particular purpose or merchantability, nor are any such representations implied with respect to the material set forth herein, and the publisher takes no responsibility with respect to such material. The publisher shall not be liable for any special, consequential, or exemplary damages resulting, in whole or in part, from the readers' use of, or reliance upon, this material.

Cover Photo: Larry Lefever from Grant Heilman

Delmar Staff
Publisher: Susan Simpfenderfer
Acquisitions Editor: Cathy L. Esperti
Developmental Editor: Andrea Edwards Myers
Project Editor: Eugenia L. Orlandi
Production Manager: Wendy A. Troeger
Production Editor: Carolyn Miller
Marketing Manager: Katherine M. Slezak

Printed in the United States of America

For more information, contact:
Delmar Publishers
5 Maxwell Drive-P.O. Box 8007
Clifton Park, NY 12065-8007

International Thomson Publishing Europe
Berkshire House 168–173
High Holborn
London, WC1V 7AA
England

Thomas Nelson Australia
102 Dodds Street
South Melbourne, 3205
Victoria, Australia

Nelson Canada
1120 Birchmount Road
Scarborough, Ontario
Canada M1K 5G4

International Thomson Editores
Campos Eliseos 385, Piso 7
Col Polanco
11560 Mexico D F Mexico

International Thomson Publishing Gmbh
Königswinterer Strasse 418
53227 Bonn
Germany

International Thomson Publishing Asia
221 Henderson Road
#05–10 Henderson Building
Singapore 0315

International Thomson Publishing – Japan
Hirakawacho Kyowa Building, 3F
2-2-1 Hirakawacho
Chiyoda-ku, 102 Tokyo
Japan

5 6 7 8 9 10 XXX

Library of Congress Cataloging-in-Publication Data
Mackenzie, Stephen Alexander.
Equine safety / Stephen A. Mackenzie.
p. cm.
Includes index.

1. Horsemanship—Safety measures. 2. Horses—Safety measures.
I. Title.
SF309.M244 1998
798.2'3'0289—DC21 97-7257
CIP

ISBN-13: 978-0-8273-7231-3
ISBN-10: 0-8273-7231-0

Dedication

To Ray Smith — the finest farrier of our time

Contents

Preface

Safety is one of the most frequently used skills in the repertoire of the person who works with horses. Though other topics are important, they cannot compete with safety. Consider the fact that you do not use your knowledge of equitation, driving, medicine, nutrition, breeding, stable management, economics, or personnel management every time you approach a horse. But for the rest of your life, every time you approach a horse, your safety skills will be called upon—every single time. Clearly it is a subject worth studying.

Many equine instructors grew up around horses and learned the hard way what not to do around them. Pain is indeed an efficient teacher, but you must survive the experience to reap the benefits. Each year there are reports of people who did not survive encounters with horses; some people technically survived such encounters but suffered permanent physical damage. If your concern is to learn how to work around horses efficiently, your first goal should be to learn how to work around them safely. You cannot help horses or work with them from a hospital bed. This book approaches safety in a systematic manner, hoping to decrease the time required to teach safety to beginners while preventing injury at the same time.

This book presents situations that are common to all who work with horses. Chapter 1 covers basic groundwork. Chapter 2 moves on to the basic restraints used to encourage horses to behave safely. Chapter 3 explains how to safely catch, lead, and tie a horse. After you have mastered these basics, proceed to Chapter 4 to learn how to apply them specifically to stalls, paddocks, and pastures, because those who work with horses spend a large portion of their time turning the horses out for exercise or bringing them back in. Chapter 5 discusses basic lungeing and ground driving, common procedures in many training systems. Failure of tack and equipment is also a common problem, discussed in Chapter 6. But the topics of riding and driving are left out. There are so many instructors available in these subjects, each with a personally preferred approach to safety, that anything written on the subject would be less than satisfactory for the majority. The good news is that all good instructors have systematic approaches to safety while riding and driving, and the reader is encouraged to spend time under the tutelage of such a person, paying particular attention to their concerns regarding safety. Finally, the book deals with safe trailering (Chapter 7). Transporting horses is such a common procedure these days, and the techniques used by different workers are so similar, that a chapter on the subject seemed warranted.

We all know that books have limitations. One that affects this book is that they cannot always teach wisdom. This is unfortunate, because wisdom and luck are what actually keep you safe around horses. Try to develop wisdom yourself, and stay away from those who disdain it. There is no safe way to work with a fool. They will get you hurt, sooner or later. It is only a matter of time. As far as cultivating wisdom is concerned, there are suggested readings. Solomon, king of ancient Israel, is reputed to have been the wisest man who ever lived. Regardless of your religious affiliations, his words, which can be found in the Old Testament book of Proverbs, are worth reading. He has much to say regarding fools and lack of wisdom, none of it positive. With regard to safety around horses, I can think of no better way to conclude than by leaving you with his words, which have proven themselves repeatedly over the centuries:

> "Forsake the foolish, and live; and go in the way of understanding."
> (Proverbs 9:6)

Good luck, and may your days with horses be long, happy and safe.

Acknowledgments

The author acknowledges the assistance of several people. Ray Smith, of Saratoga Springs, New York, who first convinced him that taking unnecessary risks is simply bad business, and who introduced him to some of the techniques described in this book. Ray Whelihan, Becky Kitts, and Dr. E. Lynn Dunn who provided important information in the chapter on tack. His wife Laurie, who has supported his work with horses and writing in many ways, and his daughter Dawn, who assisted with many of the photographs. The author and Delmar Publishers also thank the reviewers who provided valuable content expertise.

Barbara Jensen
After Hours Farms
Clifton Park, New York

Rick Parker, Ph.D.
College of Southern Idaho
Twin Falls, Idaho

Don R. Topliff, Ph.D.
Oklahoma State University
Stillwater, Oklahoma

The Basics for Groundwork

Protective Gear

Footwear

In all phases of life, it is important to have the correct equipment for projects. Safety is no exception. This begins with what is worn. Take footwear, for instance. Show me someone who works with horses and has never been stepped on by a horse and I will show you someone who has not yet spent much time actually handling horses from the ground. Being stepped on is just as much a part of life for the ground worker as falling off is for the rider, maybe more. When a large, heavy animal steps on a person's foot, it is best for that person to be wearing substantial boots. There are, of course, different types of boots. Rubber boots keep the feet dry but do not offer much support or protection from the weight of a horse. Leather and its synthetic imitations usually give the most support and protection. The choice of steel toes is a controversial one. Some people swear by them, but others avoid them. I know a man who, while wearing steel-toed boots, had a horse step on the edge of the steel cap, back away from the toes where it is thin. The steel bent, trapping his foot in the boot. Not only did he have broken bones, but he couldn't get his boot off. Though he still wears boots, no amount of money can make him wear steel-toed boots around horses anymore. This topic remains one of those personal choices that those who work with horses enjoy arguing about, but some type of boots should be worn when working around horses.

Gloves

Leather or synthetic gloves can also be quite useful. Occasionally a horse will break away from you, ripping the lead line or rope through your hands as it does so. This can cause nasty burns, especially when using nylon leads. Gloves will prevent these burns and save the flesh on your hands when you get tangled up in other types of ropes and leads.

Eye Protection

It is not unusual to get mud and dirt kicked up into your face when working horses outside in wet conditions. Goggles or other forms of eye protection should be used under such conditions. Sometimes they are necessary under dry conditions if there is enough dirt being kicked around. It is not safe either to have foreign material in the eye or to be temporarily blind and unable to see what the horse is doing.

Accessory Wear

Belts

Strong belts can even be advantageous. Over the years, they have been used as emergency leads, ropes, and whips to handle situations where there are more horses than leads or equipment. It is often not safe to have some horses controlled and others running around causing trouble. In this sort of situation, there is invariably a shortage of halters or leads. Belts should not be used regularly as leads, but in desparate situations, they have prevented many accidents and injuries when used by competent people.

Jewelry

Jewelry should be conservative when working around horses. Large earrings, especially hoops in pierced ears, lips, or eyebrows, can catch on chain crossties and other items causing a good deal of pain when they are improperly removed. One year I even had a student show up for a horse training class with a pierced navel. Try that when you are bellying up (lying on your belly) on a green horse that moves around a lot. Although I never saw the navel ring, I understand that it did not last long. Necklaces should be thin enough to break easily in order to prevent being accidentally strangled. Bracelets can also catch on things. Rings should not be large enough to touch young horses when you do not intend to. If you are working with horses whose feet have long clinches to hold shoes and extra pads on, or if you are a farrier, you may want to consider not wearing rings at all. If one of the long clinches catches under a ring and the horse spooks, yanking the foot away, it could do major damage to the finger involved. Long hair should be worn in such a way as to prevent it interfering with vision when you are working on the horse's feet and so that it will not catch on objects. If you spend some time, I'm sure you can add ideas to this list. There may be exceptions to what is stated here, but in most cases you will find that conservative jewelry is safer than the more elaborate types.

Position

Where you stand next to the horse determines what the horse is physically capable of doing to you. As such it is a major consideration for the safety of the handler. Over the years, I have found it useful to have beginners divide the space around the horse's body into what we now call safety zones (see Figure 1-1) and list what the horse can do to them in the different zones. At the end of the exercise, it is obvious to even the most basic beginner where the best places to be are when handling a horse.

Zone A

If we look at the horse's body from overhead, the first zone, labeled Zone A for simplicity, is directly in front of the horse. If you are a worker standing in Zone A, the

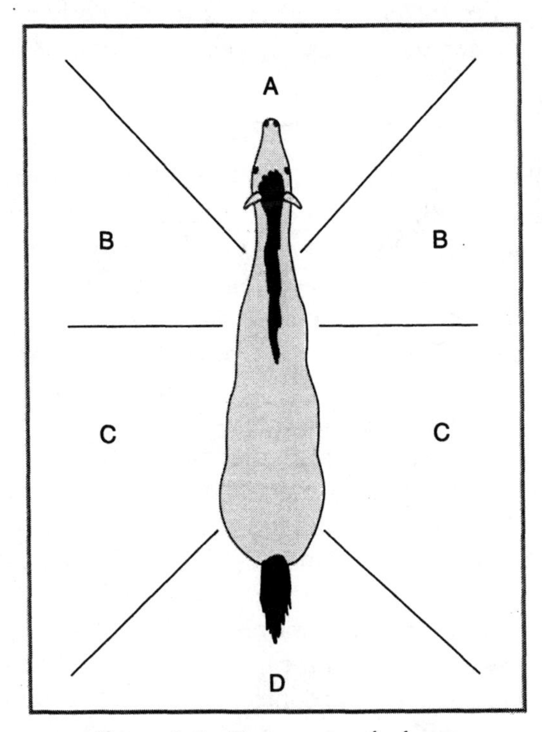

Figure 1-1 Zones around a horse

horse may run you over by bolting forward (which frequently causes major medical problems), strike you with the front feet, rear up and come down on top of you, bite you, or club you by swinging the head around on the end of that nice long neck. None of these actions is healthy for a human. Zone A is clearly not a good place to position yourself unless other options are denied you.

Zone B

The next zone, labeled Zone B, is next to the horse's shoulder. Although the horse is physically capable of reaching you in Zone B, it cannot do much efficiently or without your being able to see the problem coming. The only thing the horse can do to you efficiently in Zone B is step on your feet (at which they are quite good). Though this will cause some damage and is quite annoying, the damage is not nearly as serious to your health as that caused by the horse's actions in the other zones.

Zone C

Zone C extends from the rear of the horse's shoulder to a line swept back from the hind leg. If you are standing in Zone C, the horse can crush you up against a wall or other solid object; kick you in the head with the hind foot if you are bent over working on the front feet or reaching under the belly for the girth, and so on; lay down on you if you are working under the belly; cow kick you (a semicircular sideways motion of the hind leg); or whip you in the face with the tail (particularly painful when the horse hits you in the eye or when the tail is a large mass of burdocks).

Zone D

Zone D is essentially to the rear of the hind legs. When a human stands in this zone, he or she can be kicked with power. This can do major damage to a human, and in many cases the damage proved to be fatal. Consequently Zone D should be avoided when possible, and great care should be taken when you must work there. I often tell my students that "D" stands for "Dumb." If you must pass behind the rear end of a horse, there are two intelligent approaches. One is to be so far away from the horse that it cannot reach you with the hind feet. If you use this technique, please remember that horses' legs are usually longer than you think. The other approach is to speak first or pull down on the dock of the tail to ensure that you do not surprise the horse, and then pass so close behind the horse that your side is in contact with the buttocks. Being this close will jam any kick the horse decides to send your way. You will still feel pain from the kick, but it will not have enough room to pick up full speed before it makes contact with you, thus decreasing its power. Whatever you do, resist the temptation to push away from the horse with your arm. Being one arm's length away from the rear end of a horse keeps you well within range of the hind feet and yet gives the feet enough room to pick up full speed before they make contact with you. You have just given the horse perfect conditions to kick with power—not a smart thing to do.

Staying in the "Safe Zone"

An examination of the zones quickly reveals that Zone B has fewer problems associated with it. For this reason it is considered the safest place to position yourself when handling a horse. The trick is to stay in Zone B, guard against being stepped on, and wear good boots. If you must leave Zone B to complete your work, it is usually safer to stay right next to the horse to jam any kicks that might be aimed at you. The problem, of course, is that horses move. As they move, Zone B moves with them. Frequently staying safe is a matter of competing for position. The horse will spin and turn, trying to get you out of Zone B, and you must keep moving quickly to stay in it. When a horse acts up dangerously, it is often best to turn it in a small circle as sharply as possible. This keeps it from being able to face straight away from you, kick, or pull away by bolting forward (which allows it to use its most powerful muscles against you). The question is, should you turn to the right or to the left? When evaluating techniques to determine which is safer, the general principle is that whichever technique keeps you in Zone B the greatest amount of time is the safest one. In this case, turning to the left allows you to stay in Zone B throughout the maneuver (see Figure 1-2). Turning to the right, a practice encouraged in the show ring and in youth organizations, is difficult to do with a horse that is resisting you. If the horse does not decide to yield to you, you must push the head over as best you can, stepping into Zone A for maximum effect (see Figure 1-3). As soon as you do so, you expose yourself to all the safety problems listed earlier under Zone A. Therefore, with a difficult or dangerous horse it is best to turn it to the left whenever possible. At first glance it may seem that show procedures and youth organizations need to change their emphasis, but I do not believe that to be the case. Turning to the right in the show ring and in youth organizations makes good sense, due to the nature of the horses involved. By the time you are in the show ring,

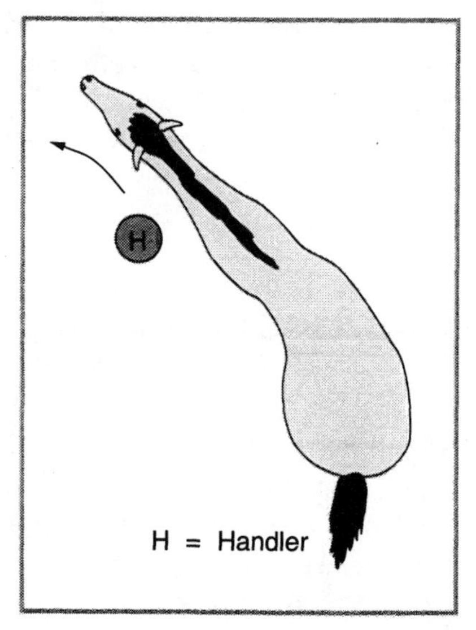

Figure 1-2 Staying in Zone B by turning to the left

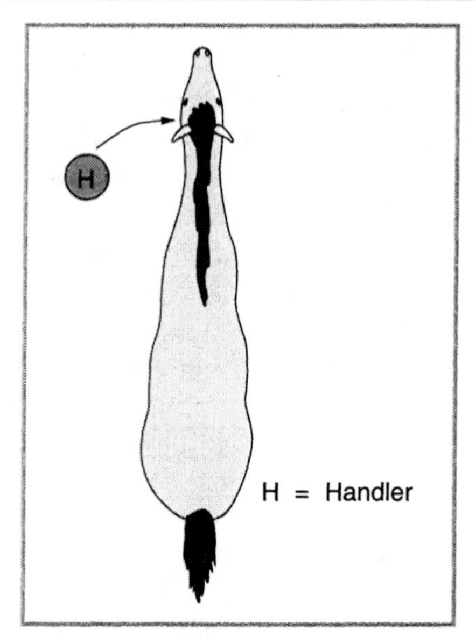

Figure 1-3 Entering Zone A by turning to the right

your horse should be well mannered enough to yield to your leading. People in youth groups usually encourage their students to work with kindly horses who do not intend to hurt people. When dealing with such animals, the biggest problem you face routinely is that the horse might accidentally step on you as you lead it. Zone B has the weakness of placing your feet close to those of the horse, and turning to the right with a compliant horse makes it harder for the horse to step on you. Because young people and beginners tend to get discouraged if the horse repeatedly steps on them, it is good for the horse business to match them with compliant horses and have them turn to the right whenever possible. However, when you leave the realm of benevolent or well-behaved horses, your safety problems are much more serious than having your feet stepped on. You should school yourself to turn to the left when things get difficult with young, fractious, or dangerous horses. Holding up the elbow of the arm nearest the horse will allow you to keep your feet from being stepped on. To turn a horse to the right when it is misbehaving is a dangerous thing to do.

WORKING WITH AND AROUND HORSES' FEET

Front Feet

One of the most common procedures involved with working with horses is picking up their feet. Before doing this, be sure that you have enough free space to work in. Do not get thrown into walls or objects when there is another, larger area you could have chosen to work in. How you proceed depends upon whether you are picking up a front or hind foot. If you are picking up a front foot, you first position yourself in Zone B next to the shoulder. Reaching down the back of the leg with the hand nearest the horse, squeeze the tendons behind the pastern bone, just below the horse's knee (see Figure 1-4). While you are doing this, be sure not to lean so far toward the hind feet that you give the horse the chance to kick you in the head with a hind foot. Most horses will eventually respond to this squeezing of the tendons by lifting the foot off the ground. Sometimes it helps to lean into the horse, shifting its weight to the opposite side before squeezing the tendons. As soon as the foot comes up, grasp the point of the toe to control it (see Figure 1-5). If the horse doesn't think that you are confident or in control, it is more likely to take the foot back, so grasp the foot firmly and confidently. In the event that the horse does take the foot back, do not hold on long enough for the hand holding the toe to be pulled down underneath the foot. If two hands are needed for working, step across the foot with the leg nearest the horse and grip the horse's foot between your legs just above your knees (see Figure 1-6). It helps if you keep your knees slightly bent and your feet slightly pigeon-toed. Should the horse act up while you are in this position, release the foot with your knees, and straighten up, lifting the inside leg (the one nearest the horse) while pushing away with the inside arm (see Figure 1-7). Keep your weight on your outside foot and pull the inside leg up and away, do not plant both feet and try to jump clear. Trying to jump is much slower and keeps your inside foot closer to the horse's legs for a longer period of time. This increases the likelihood that you will get stepped on or tangled up with the horse. Standing up on your outside foot and pushing away with your inside arm is the fastest way to get clear.

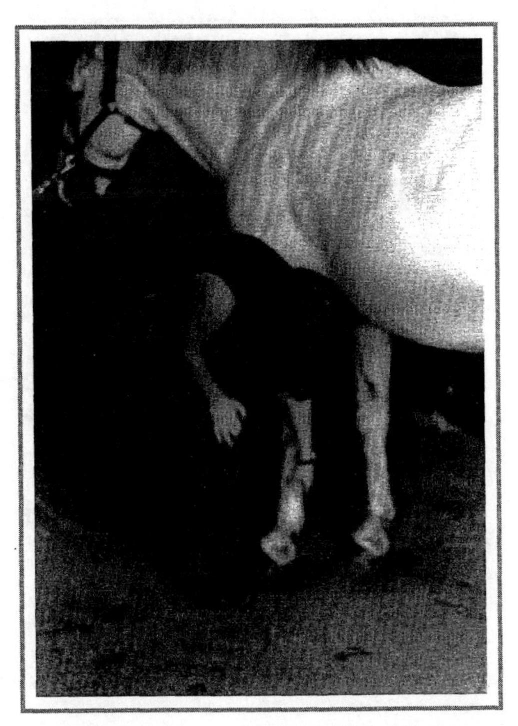

Figure 1-4 Picking up a front foot

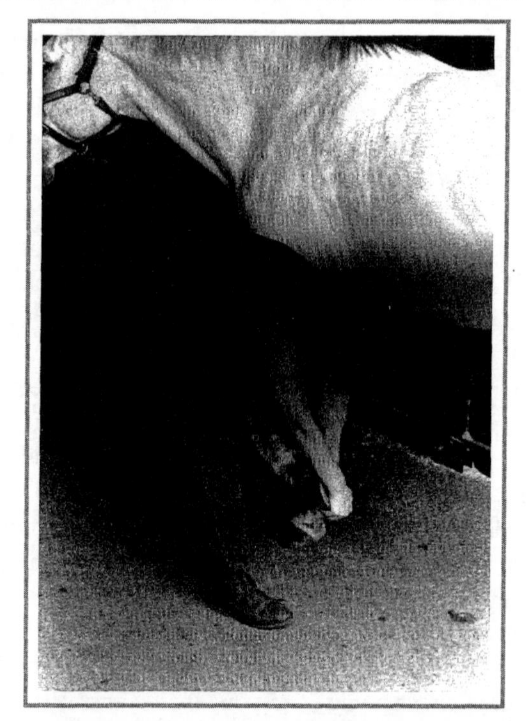

Figure 1-5 Grasping the point of the toe

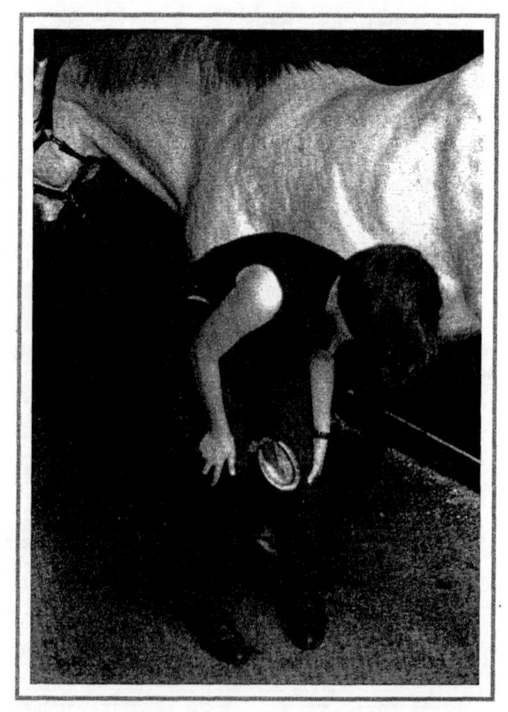

Figure 1-6 Keeping the hands free on the front foot

Figure 1-7 The front foot escape

Hind Feet

If you are picking up a hind foot, there are two popular methods to choose from. In the first, you position yourself next to the horse's hip facing the tail with the hand nearest the horse touching the hip in case you need to push away to the side. With the remaining hand, you slide down the rear side of the leg and grasp the middle or lower section of the cannon bone (see Figure 1-8). Firmly pulling the leg forward toward the horse's head will cause the joints of the leg to bend in such a way as to raise the foot off the ground (see Figure 1-9). Do not try to pick the foot straight up, it is much too difficult. Pulling the leg forward will allow the leg joints to do the work for you. As soon as the foot clears the ground, grasp the point of the toe with your free hand to control the leg (see Figure 1-10). If two free hands are necessary for working, place the leg of the horse over the thigh of the inside leg, blocking the toe of the hoof with the knee of your outside leg. Placing the armpit of your inside arm over the hock joint will allow you to control the leg without using your hands. They are then free for whatever work may be necessary (see Figure 1-11). This position takes some practice to master, but it is well worth the effort. It helps if you face across the horse's body axis at about forty-five degrees and spread your feet slightly, keeping them slightly pigeon-toed (see Figure 1-12). Any good farrier (horseshoer) can coach you on this.

The other option for picking up the hind feet begins by positioning yourself as you do to begin the method just described. This time, however, you reach down the leg with

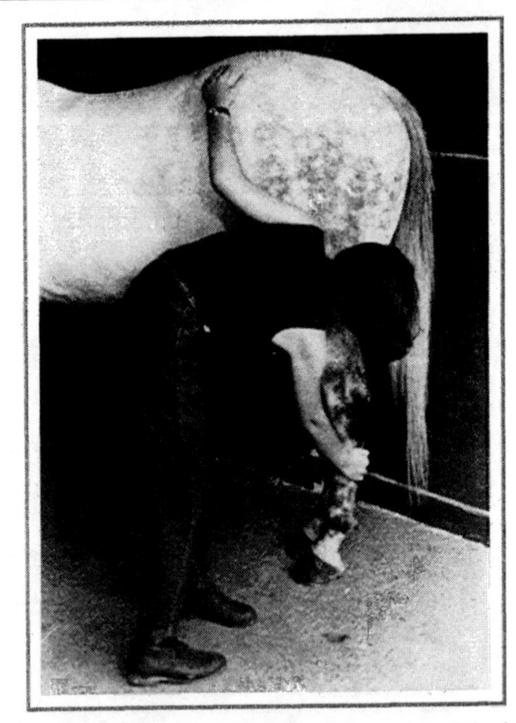

Figure 1-8 Keeping a hand on the point of the hip

Figure 1-9 Pulling forward to raise the foot

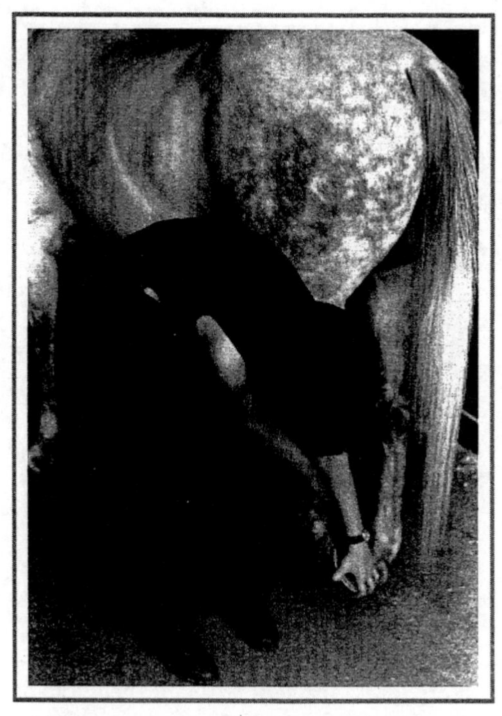

Figure 1-10 Grasping the point of the toe

Figure 1-11 Keeping the hands free on the hind foot

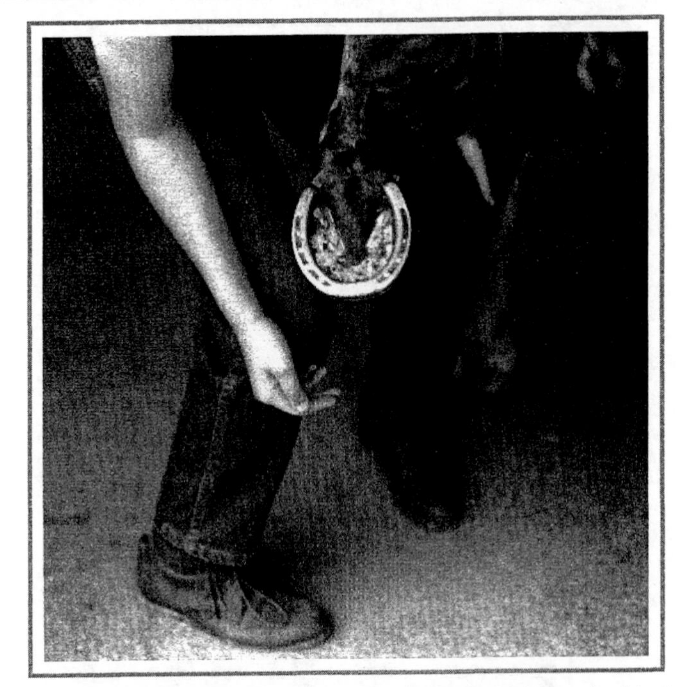

Figure 1-12 Body angle and foot position when holding the hind foot

the inside hand (the one nearest the horse) (see Figure 1-13). If the horse acts up, you use the elbow and forearm of the inside arm to push away to the side. When you reach the lower section of the cannon bone, you pull the leg forward toward the head as in the previous method (see Figure 1-14), grasping the point of the toe as it gets high enough to do so (see Figure 1-15). You can then hold the foot as described for the first method.

The Handler

Handling the horse when someone else is picking up the feet or working on it is an important job. When the horse acts up, you need to give the worker the best possible chance of surviving unscathed. This is often accomplished by maneuvering the horse to put the worker in the best possible position. The key to this maneuvering is to remember that when turning a horse from the ground, the rear end usually moves the opposite direction from the head. For instance, if you pull the horse's head to its left, the rear end will move to the horse's right, and vice versa (see Figures 1-16 and 1-17). The other important item to remember is that it is safer and far more efficient to pull a resisting horse's head in your direction than it is to try to push it away from you. Pulling the head keeps you in Zone B, allowing you to put the weight of your body into the effort; pushing the head tempts you to step into Zone A and pits the strength of your arms against that of the horse's neck. With these ideas in mind, it should become clear that where you stand when handling the horse is important for the safety of the person working on it.

Figure 1-13 Using the inside hand to pick up a hind foot

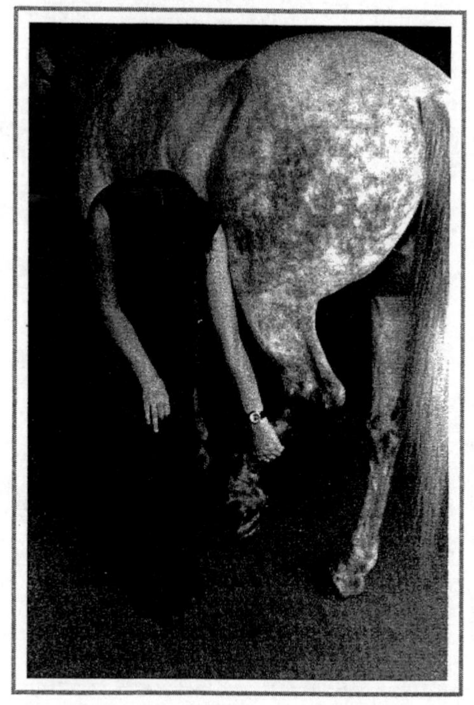

Figure 1-14 Pulling forward with the inside hand to raise the hind foot

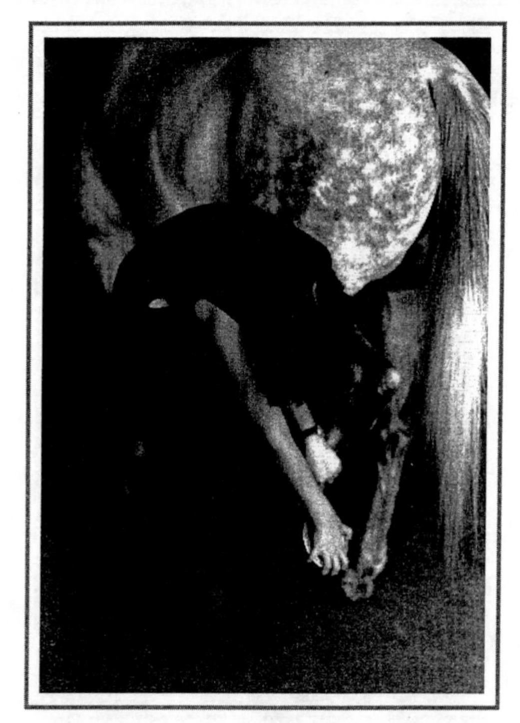

Figure 1-15 Grasping the point of the toe

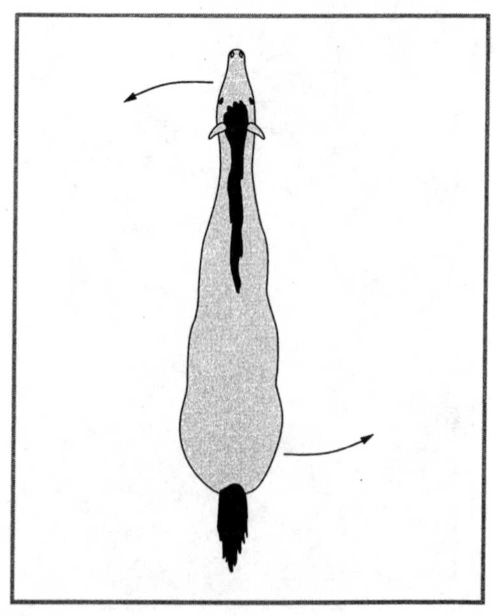

Figure 1-16 When the horse's front end moves to the left, its rear end moves to the right.

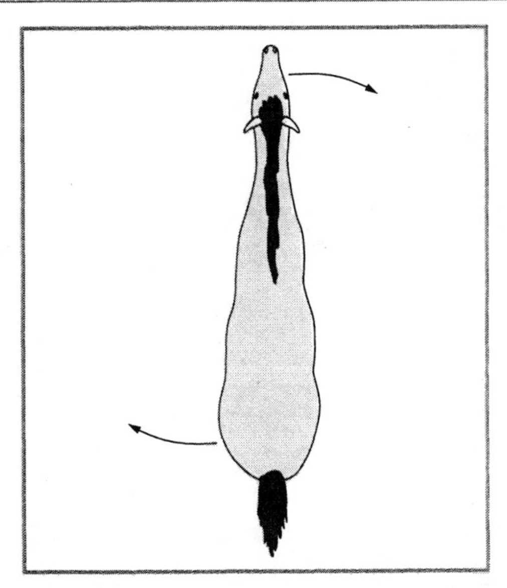

Figure 1-17 When the horse's front end moves to the right, its rear end moves to the left.

Where the handler is positioned is determined by whether the worker is on the hind end of the horse or the front. Although different approaches work, it is important that the handler and the worker agree on which approach they will use and that both be experienced. When the worker is on the front end of the horse, there are two major approaches: one that protects him or her from the front end of the horse, and the other from the hind end. If you are more concerned with protecting the worker from the front end of the horse, the handler positions himself on the opposite side from the worker. When the horse acts up, the worker escapes toward the front end of the horse while the handler pulls the head enough to get one sideways step from the horse (see Figure 1-18). More than one step is not desirable, because that spins the rear end of the horse around to threaten the worker. However, when done correctly, this is the fastest way of separating the worker from the front end of the horse. Those of us who have had the front end of a rearing horse come down on top of us tend to choose this approach.

On the other hand, some workers have had the handler lose control while they were working on the front end of the horse, allowing the rear end to spin around. They tend to choose the next method, in which the handler stands on the same side of the horse as the worker. When the horse acts up, the worker escapes toward the side or rear of the horse while the handler pulls the head, moving the rear end away from them (see Figure 1-19). Both of these approaches work if the two people involved coordinate their actions properly. For this reason, people should practice escape drills as described here and agree upon which approach they will use before beginning work on a horse.

When the worker is on the hind end of the horse, most people agree that the handler should stand on the same side of the horse as the worker. Then when the horse acts up, the handler can pull the horse's head and move the rear end of the horse away from the worker (see Figure 1-20).

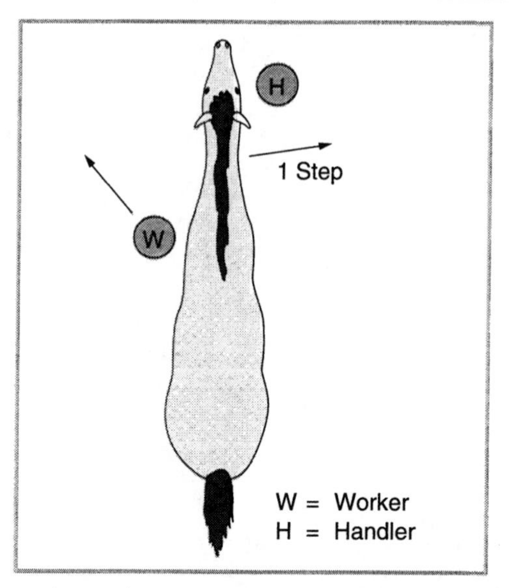

Figure 1-18 The worker escapes toward the front end of the horse.

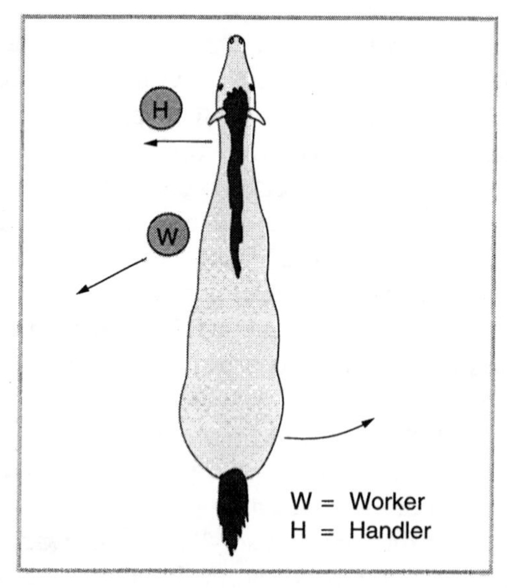

Figure 1-19 The worker escapes toward the rear end of the horse.

Sometimes you must work with a horse in a confined area, such as its box stall. In this case you have less ability to maneuver, so many people prefer to have the handler on the same side of the horse as the worker. This simplifies things for the handler, because there is less to think about. If there is danger of the horse acting up, a long-handled twitch (see Chapter 2 for an explanation of twitches) or tranquilizers may be useful (see Figure 1-21).

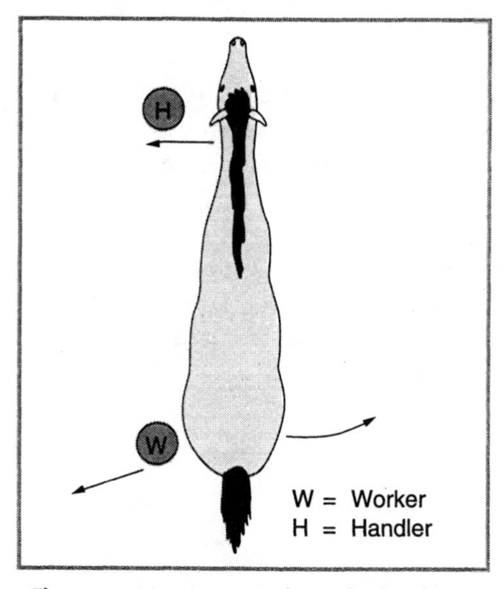

Figure 1-20 Escaping from the hind legs

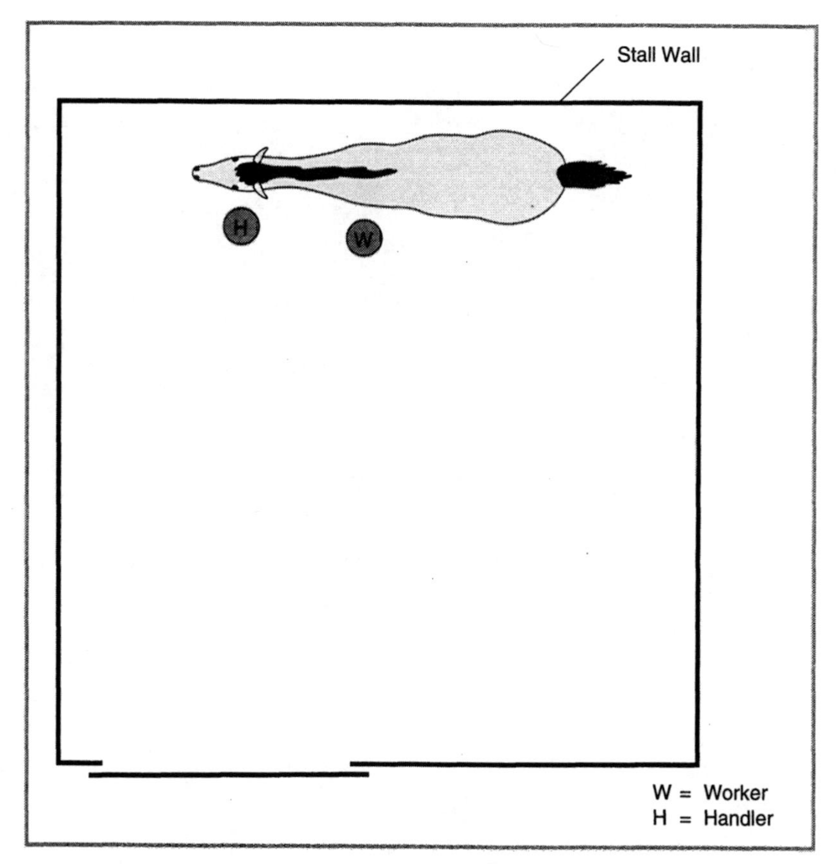

Figure 1-21 Working on a horse in a box stall

SUMMARY

Regardless of the approach used, it is valuable for the handler to be able to distract and entertain the horse. Many horses give us trouble because they have nothing else engaging their minds. When they are properly entertained by the handler, they give the worker much less trouble.

As you can imagine, the skill and judgment of the handler are critical items when you are working on a horse. If the person you are working with is foolish, no approach will offer you much chance of staying safe. There is no good way to work with fools. Foolish people inevitably find ways to get you into trouble, and they are usually no good at getting you back out. Don't work with them if you don't have to, and don't trust them when you do.

Basic Restraints

THE HALTER AND LEAD

The handler's ability to restrain and control the horse is often the critical factor in keeping the worker safe. Methods of restraining horses are numerous, and it would not be possible to cover all of them. This chapter focuses on several widely employed methods, keeping in mind that most people experienced in working with horses have favorite and sometimes unique methods of their own.

The simplest and most universal method of restraint is the halter. When combined with a lead shank equipped with a section of chain on the snap end (often referred to as a chain lead shank), the halter can offer different levels of restraint. The lowest level of control is offered by simply snapping the lead shank into the halter ring located under the horse's chin (see Figure 2-1). In this manner, the handler can hold the horse's head still or pull it in the desired direction if the horse does not resist. Many people hold the halter itself to better control the head's movement, but this exposes their fingers to being broken should the horse move suddenly (which they often do). It is considered best to hold the lead shank ten or twelve inches away from the head, unless the horse is a known biter. The excess lead should never be tied to the handler's body or wrapped around the hand (often referred to as the suicide grip). Horses can move suddenly, and it is important for the handler to be able to separate from the horse quickly when necessary. The excess lead can be coiled or folded and held in the hand farther away from the halter.

The next higher level of control is offered by passing the chain lead shank through the halter ring on the horse's left side (referred to as the "near" side), crossing it over

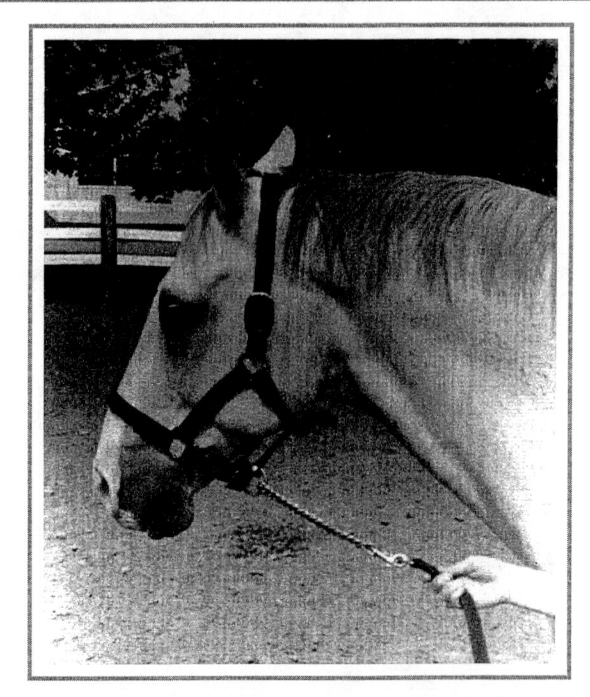

Figure 2-1 The lead shank snapped to a halter ring

the nose, and snapping it into the halter ring on the horse's right side (referred to as the "off" side), sometimes running the chain up to the ring higher up the head if the chain is long enough. This procedure is often called "putting the chain over the horse's nose" (see Figure 2-2). If the horse acts up, the handler can yank down sharply on the lead, causing a certain amount of pain to the horse's nose. This will often allow the handler to reestablish control, and many horses behave well simply because the chain is there. With these horses it is not necessary to use the chain, but if it is not over the nose, the horse may take advantage of the handler. For this reason, many people put the chain over the nose as a routine practice, unless there is some reason not to. Care must be taken to not let the chain slip down too far on the nose, because it might damage the cartilage on the end of the nose when the lead is yanked. For this reason, many people wrap the chain once over the nose band of the halter as they pass the chain over the nose. This prevents the chain from slipping down far enough to hurt the cartilage.

Putting the chain under the chin offers the same level of control as putting it over the nose. In this case, the chain is passed through the halter ring on the near side (the horse's left side) and under the chin, and the snap is connected to the halter ring on the off side (the horse's right side), running the chain up to the ring higher up the head if it is long enough (see Figure 2-3). This method is useful for horses who object to having the chain put over the nose for either physical or mental reasons. Many people feel that putting the chain under the chin teaches the horse to throw its head up too much. This is not true. What is true is that any time a horse feels pain in the head, it raises its head. But this happens whether the chain is over the nose or under the chin.

Figure 2-2 The chain over the horse's nose

Figure 2-3 The chain under the horse's chin

A higher level of control is obtained by putting the chain in the horse's mouth. This is a technique requiring good judgment, because the chain has more chance of causing physical damage to the horse. One way to put it in the mouth is by passing the chain through the halter ring on the near side, under the chin, and snapping it to the halter ring on the off side. The chain is then pulled out with the left hand to create extra length. Holding this extra length with the left hand and inserting the thumb of the left hand into the horse's mouth at the corner of the lips where there are no teeth, it is possible to pull the chain into the mouth as the horse opens it in response to the insertion of the thumb (see Figure 2-4). Yanking on a chain in the mouth creates more pain than the previous methods and therefore allows the handler to control more-difficult horses, but great care must used. It is possible to damage the horse's tongue if the chain is used too severely. As with other methods, many difficult horses behave well if they know that the chain is in place, whereas they act up if it is not. With these horses it is wise to put the chain in the mouth as a routine procedure, especially because it will not have to be used frequently. Perhaps the best use of the chain in the mouth is not to create pain but to distract certain horses. Some young horses move around when being worked on because they are bored or because their attention span has been exceeded. With the chain in the mouth, they have something to chew on and play with, which allows them to stand still for longer periods of time. Obviously with this type of horse, great pains should be taken not to yank on the lead accidentally.

Figure 2-4 The chain in the horse's mouth

Figure 2-5 The chain over the gums

The most severe use of the chain for restraint is to place it over the top gums in the horse's mouth. Obviously, this requires the most sound judgment of all the techniques. One way to accomplish this positioning of the chain is to feed it through the halter ring on the near side, run it under the chin, and connect the snap to the halter ring on the off side. After pulling the chain through with the left hand to create slack, the left hand can pull the chain back across the gums as the thumb of the left hand pushes the upper lip up and out of the way so that it doesn't get caught under the chain (see Figure 2-5). The chain must be held snug when in place or it will slip off the gums, but the slightest tug on the lead will create substantial amounts of pain. As with having the chain in the mouth, this method should be used with care.

The War Bridle

The previously mentioned restraint methods are essentially ways of inflicting pain in a controlled manner. There is one more that does not involve the halter, but a section of rope or line. Over the years it has come to be known as the war bridle. Actually, there are several versions of the war bridle, some of which have the appearance of tools from the Spanish Inquisition. What I describe here is what many people call the simple war bridle, which involves nothing more than a piece of rope with a loop either spliced or

Figure 2-6 A simple version of the war bridle

tied in one end. The unlooped end of the rope is passed through the loop to form a larger sliding loop about the size of the horse's head. The sliding loop is placed over the horse's head so that it fits behind the ears and in the mouth like a bit. The original small loop should be on the near side of the horse's head pointing down toward the ground (see Figure 2-6). When this is fitted properly, a yank downward on the free end of the rope will pull down on the poll and up on the corners of the mouth at the same time, causing quite a bit of pain. In many cases, this allows the handler to gain control over very difficult horses.

Basic Physical Restraints

So far the text has described using pain to restrain the horse, but quality workers do not enjoy inflicting pain. Though being prepared to use pain is necessary, one should never enjoy it. Fortunately there are other ways to restrain horses that often eliminate the need for pain. Using them, however, does not eliminate the need to have the lead shank placed properly in case the horse acts up. Always prepare for the worst, just in case.

Basic physical restraints begin with simply holding the nose down with your hand (see Figure 2-7). Often this is enough to quiet the horse and keep it from moving. When you do this, be aware that you can decrease the airflow through the nostrils if you do

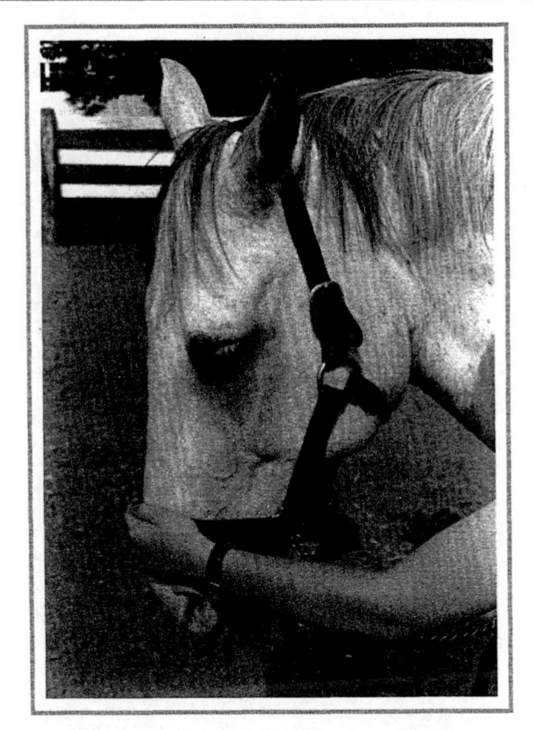

Figure 2-7 Holding the nose down

not do it properly. If the horse feels that its air supply is being compromised, it may panic and begin to fight you. Naturally, this will cause more problems than you had in the first place.

Sometimes shifting the horse's weight restrains the part you wish to work on. For instance, to work on a front foot or leg, the handler can pick up and hold the opposite front foot (see Figure 2-8). The horse can't move the foot it is standing on, which allows the worker access to the top portions of it and the leg without it moving.

Shifting the weight onto a hind foot can work just as well. A simple way to do this is to pull the tail toward the foot or leg the worker needs to access (see Figure 2-9). This places the horse's weight on that foot, making it hard for the horse to move it or kick. Picking up the opposite hind foot would work just as well. Please note that the horse can still kick from these positions, but this makes it slightly more difficult. If you motivate the horse to kick, no weight shift will thwart it. They can do amazing things when they really want to kick.

Young foals can be quite difficult to manage at times. They can bite, move quickly, and kick forcefully enough to break human kneecaps. One of the more popular restraints for them is to place a rope or arm around the neck and grasp the tail with the other hand, raising it above the level of the back (see Figure 2-10). It is even better if the handler can get the foal up against a wall. For some reason, raising the tail seems to immobilize the rear end of the foal, making life much easier for the humans.

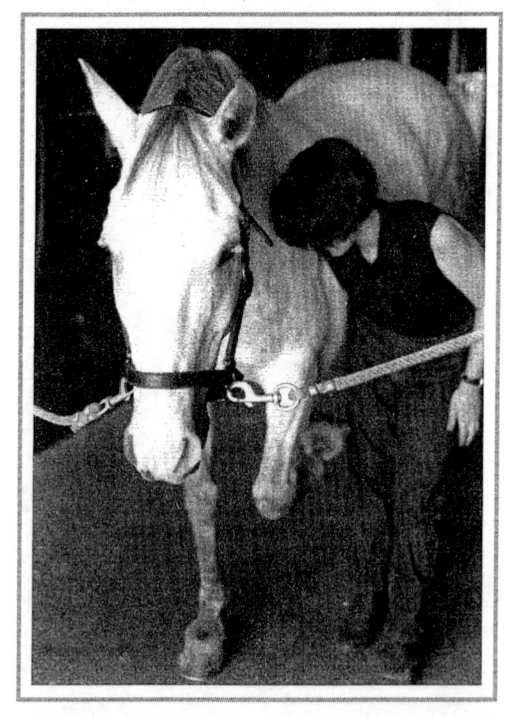

Figure 2-8 Holding up a front foot to allow work to be done on the opposite foot or leg

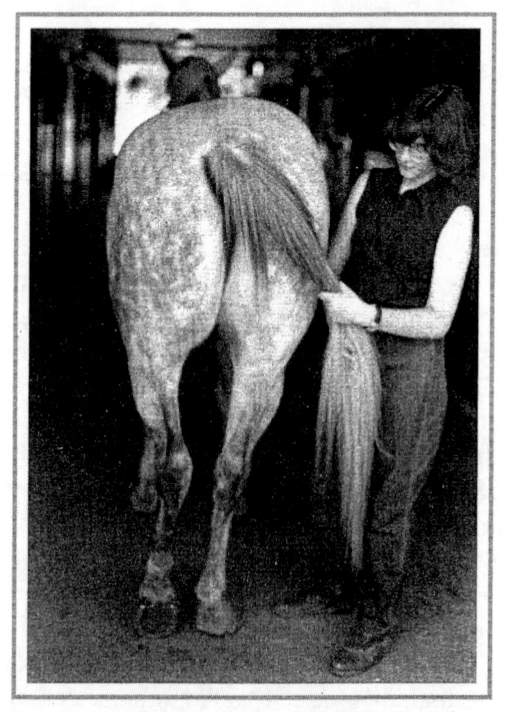

Figure 2-9 Gently pulling the tail to shift weight onto a hind foot

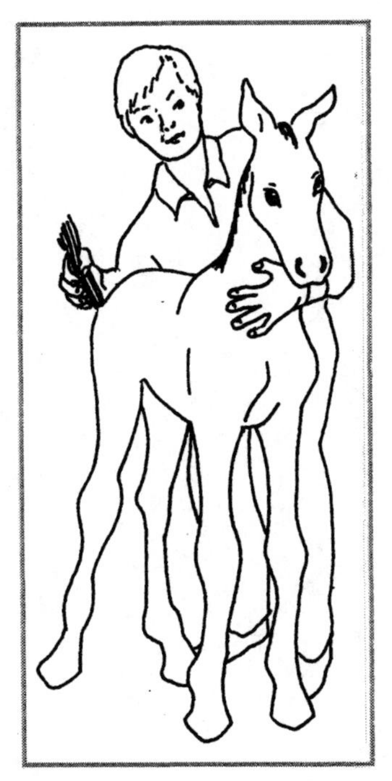

Figure 2-10 Restraining a foal

HOBBLES

Sometimes you need to work on the ground surface of the foot, and the horse is too strong for the handler to hold it up. In this case, a length of rope is useful. Do not use thin rope. Try to get the thickest rope that will do the job. The thicker the rope, the more surface area the pressure is distributed over. Thin rope concentrates the pressure in a small surface area, which can cause burning and pressure injuries. It is best to use a leather pastern strap or hobble attached to the pastern of the foot the handler wishes to raise. The rope is then tied to the D ring in the strap. If no strap is available, tie a bowline knot snugly around the pastern (see Figure 2-11). Pass the free end of the rope over the withers so that the handler can hold it on the opposite side of the horse. If the worker can induce the horse to lift the foot, the handler pulls the rope at the same time and holds the foot up by pulling the rope down over the withers (see Figure 2-12). Many people use a figure 8 hobble for this procedure, one end attached to the pastern and the other to the forearm. This allows one person to work on the horse; so if you are alone, you may have no choice but to do it this way. However, it is not as easy to release if the horse panics. Although many horses behave quite well with the hobble applied in this manner, others have been known to throw themselves over in fright, trying to free themselves from such an unyielding restriction. It is therefore safer for both horse and human to have two people and use the rope, which can be released immediately when necessary.

Figure 2-11 The bowline knot

Figure 2-12 Using a rope to hold up a front foot

Sometimes a rope is handy for the hind feet as well. It can be used to provide safety for the worker and encourage the horse to hold its foot up at the same time. It requires a third person to hold the horse or an animal that will stand well when tied. The first step is to place a pastern strap or hobble on the pastern of the foot the worker wishes to raise. The rope is then passed through it and attached to the horse's tail. This can be done by splicing or tying a large ring to the end of the rope and then braiding the ring into the hairs of the tail if they are long and strong enough. Or the rope can be tied directly into the tail using a sheet bend knot (see Figure 2-13). The free end of the rope is then held by the handler, who makes sure to stay out of range of the hind feet. When the rope is pulled, the foot will be lifted off the ground, allowing the worker to proceed (see Figure 2-14). If the horse kicks, it pulls on its own tail. Most horses soon learn that their tail feels better if they hold the foot still. The handler can adjust the tension on the tail, as the situation dictates, and can even set the foot completely down when necessary. This method offers greater safety for the worker and is a big advantage in retraining frightened horses or in initial training of difficult young horses. There are other methods that tie the hind foot up more permanently so that it cannot be put back down. Some of them offer greater safety for the worker with horses that won't let anyone touch the hind foot to apply the pastern strap. However, like the figure 8 hobble on the front leg, they are more difficult to adjust and release. Although they are valuable for people working in isolated areas without access to veterinarians and chemical restraints, or with horses known to have negative reactions to tranquilizers, horses have been known to panic and throw themselves over trying to escape from

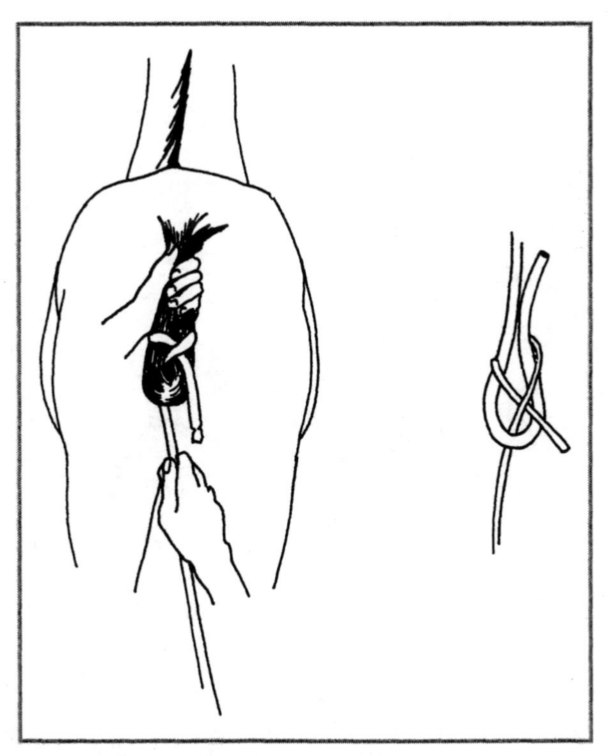

Figure 2-13 The sheetbend knot

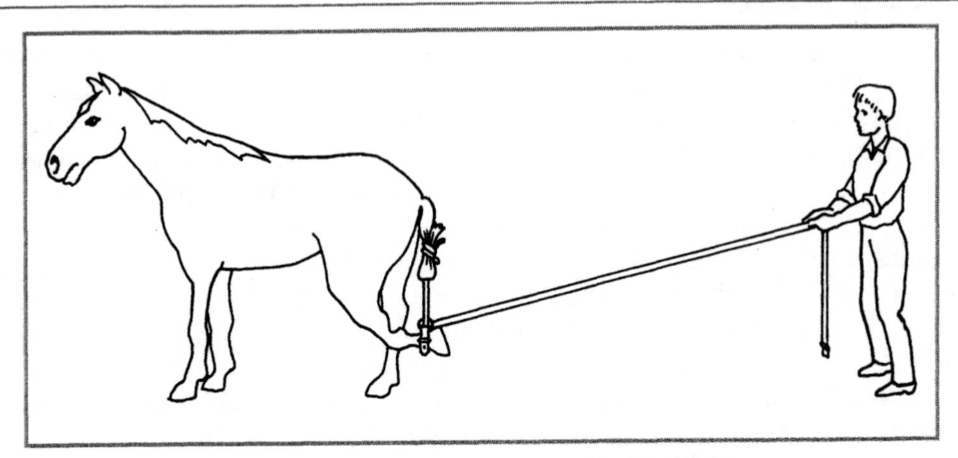

Figure 2-14 Using a rope to lift and hold a hind leg

them. This can cause injury to both horse and human. It is safer to seek the assistance of chemical restraints and use adjustable rope techniques when possible.

Special hobbles have been designed for horses that have a tendency to kick with the hind feet. Known as kicking hobbles, or breeding hobbles, they are frequently used in breeding situations where the stallion needs protection from the mare (see Figure 2-15); however, they need not be restricted to this use.

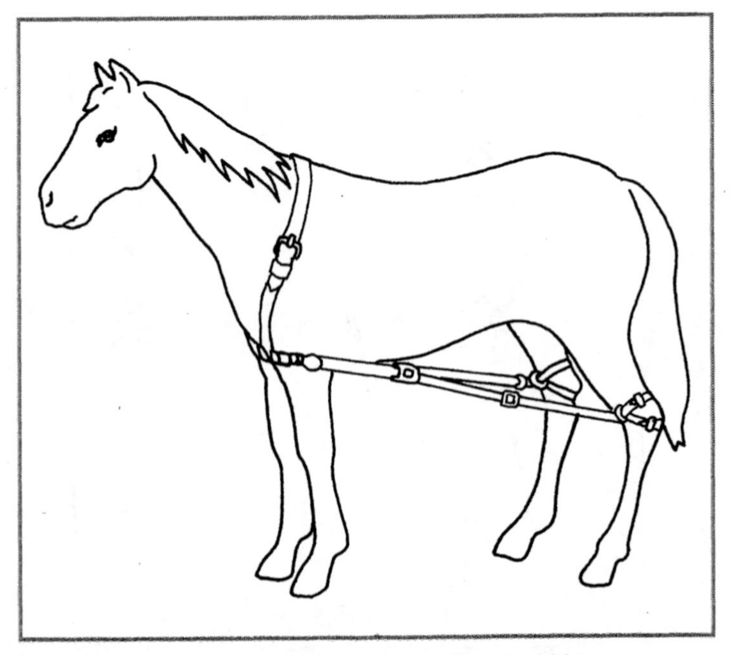

Figure 2-15 Kicking, or breeding, hobbles

Figure 2-16 Equine stocks

STOCKS

Many horses stand well in permanent structures known as stocks (see Figure 2-16). One end is usually left open, and the horse is walked into the small enclosure. After the horse is in, the open end is closed, confining the animal in so small an enclosure that it cannot move around enough to cause a safety concern. If they are trained to stand in them, horses can be worked on easily and safely. If they are not trained to stand in them, they can panic and thrash around, causing several types of injury. For this reason, tranquilizers are often used in conjunction with stocks, although it is best to train your horses to stand comfortably in them so that chemical restraints are not necessary.

MUZZLE

The prevention of kicks is a critical safety concern, but as anyone who works with horses can tell you, the bite of a horse can be very destructive. Physical restraints for the mouth include the muzzle, which is quite effective but does not allow the worker to touch the mouth itself (see Figure 2-17). When it is necessary to work on the mouth, oral speculums and gags may be employed (see Figure 2-18).

Figure 2-17 The equine muzzle

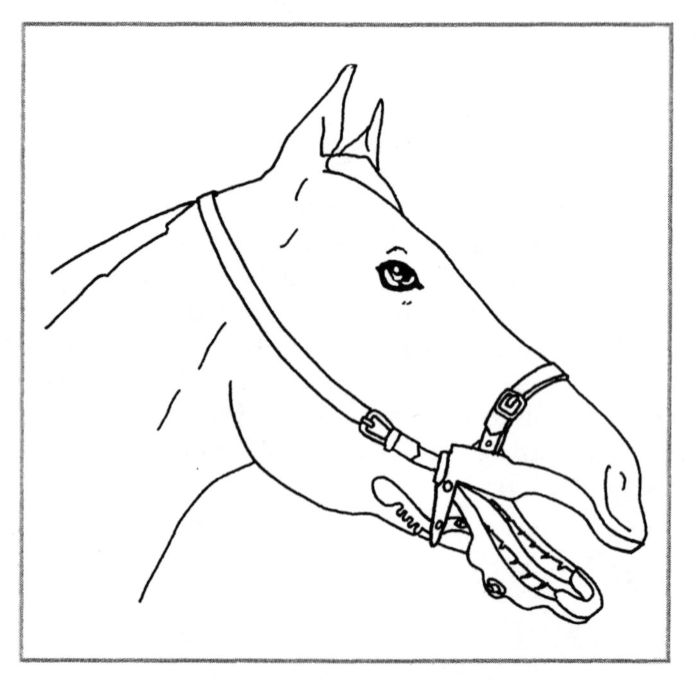

Figure 2-18 The equine speculum

TWITCH

For hundreds of years, people who have worked with horses have known how to calm unruly horses by applying what is known as a "twitch." This device is usually applied to the horse's upper lip, and over time it can almost lull some horses to sleep. For a long time it was thought that the minor pain created by the twitch was distracting the horse and focusing its attention on its nose. This is undoubtedly true, but it does not explain the calm, almost sleepy reaction of many animals. In recent years, researchers have shed some interesting light on this phenomenon. The twitch may actually be a simple form of chemical restraint. It seems that when the twitch is applied, certain chemical compounds called "endorphins" are released from the brain. These endorphins are narcotic substances that occur naturally in the body and serve many purposes, one of which is the suppression of pain. So what is happening seems to be that the twitch distracts the horse momentarily, and then the hose feels less pain and the drowsy effects of the drug application. This complicates the argument of those who think twitches to be a cruel remnant of past ages. When there is no tranquilizer or pain killer available, the twitch lowers the pain level experienced by the horse and keeps the worker safer at the same time. Its value may need to be reconsidered in light of this new evidence.

Hand Twitch

The simplest twitch is applied by the handler's hand. He or she can grasp the upper lip of the horse and simply hold on until the worker finishes (see Figure 2-19). Other people prefer to grasp the horse's skin at the shoulder, which works just as well (see

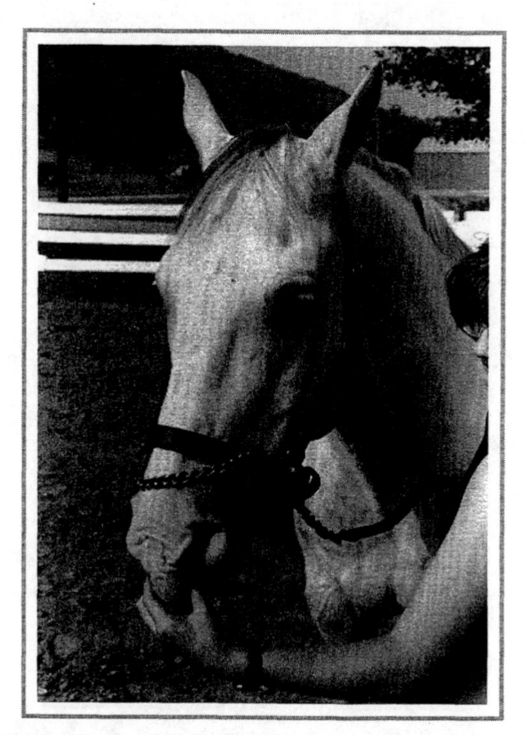

Figure 2-19 The hand twitch applied to the nose

Figure 2-20). Another version of the hand twitch is to grasp the ear of the horse (see Figure 2-21), although this should be used sparingly to avoid creating ear-sensitive horses.

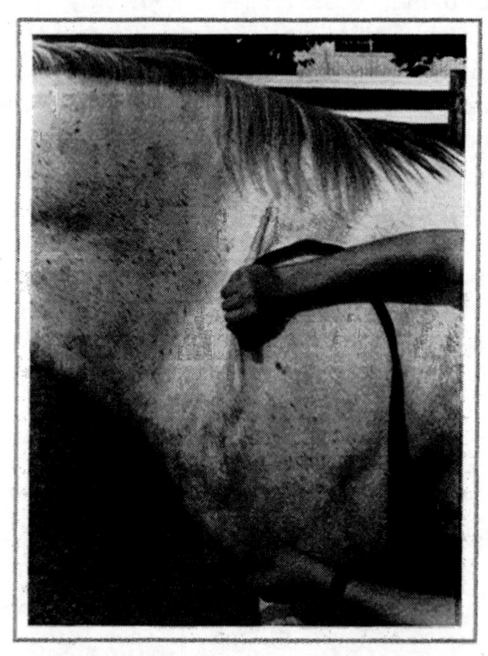

Figure 2-20 The hand twitch applied to the skin of the shoulder

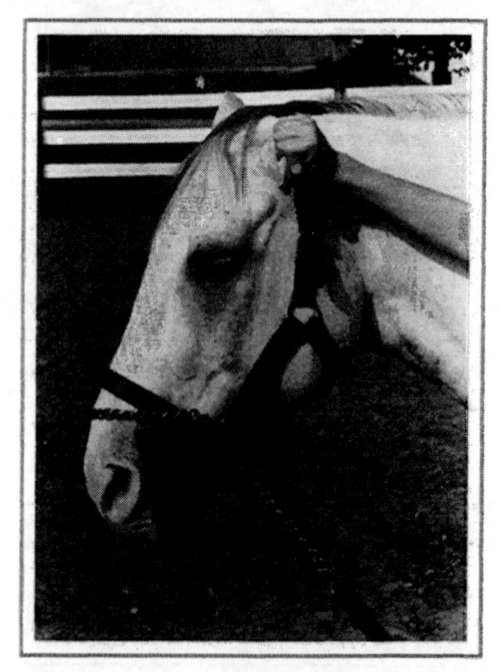

Figure 2-21 The hand twitch applied to the ear

Rope Twitch

Perhaps the most common twitch is the rope twitch, which is a loop of rope on the end of a handle of appropriate length (see Figures 2-22 and 2-23). It can be applied by holding the handle in the right hand and reaching through the loop with all but the

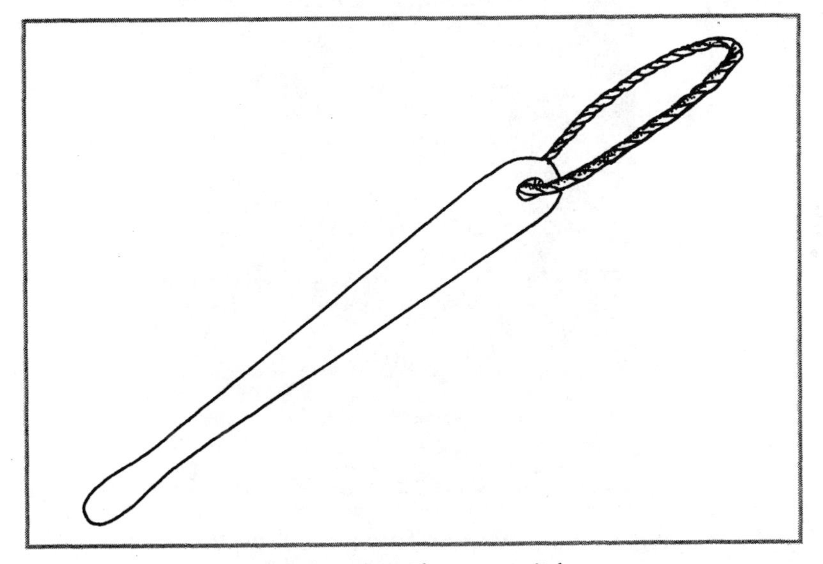

Figure 2-22 The rope twitch

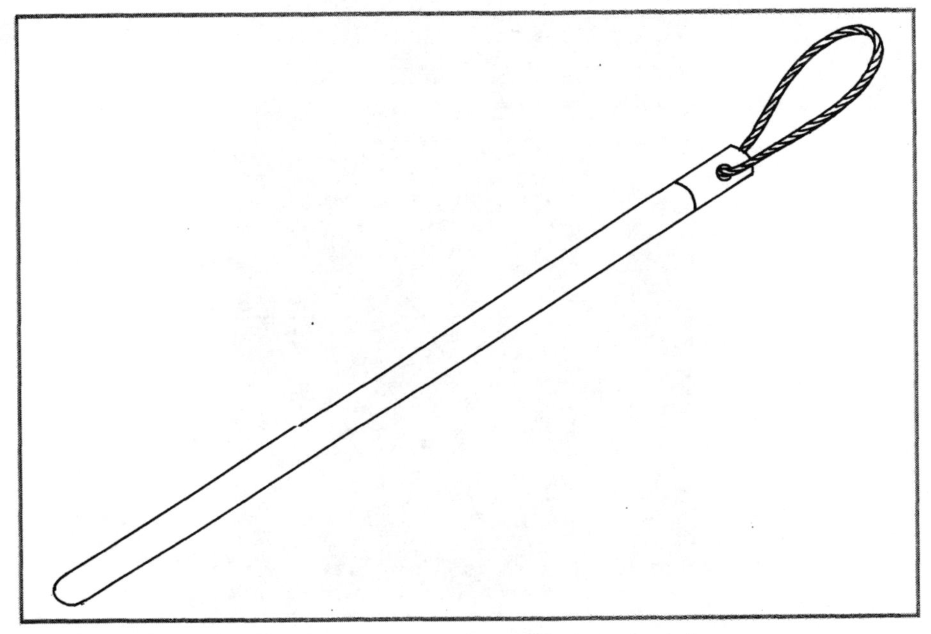

Figure 2-23 The long-handled rope twitch

small finger on the left hand (see Figure 2-24). The three fingers and thumb that are extending through the loop grasp the horse's upper lip and the rope loop is slipped off the hand so that it encircles the upper lip (see Figure 2-25). While the lip is held with the left hand, the handle is twisted until the rope loop itself is holding tight to the upper

Figure 2-24 Rope positioned for applying the twitch

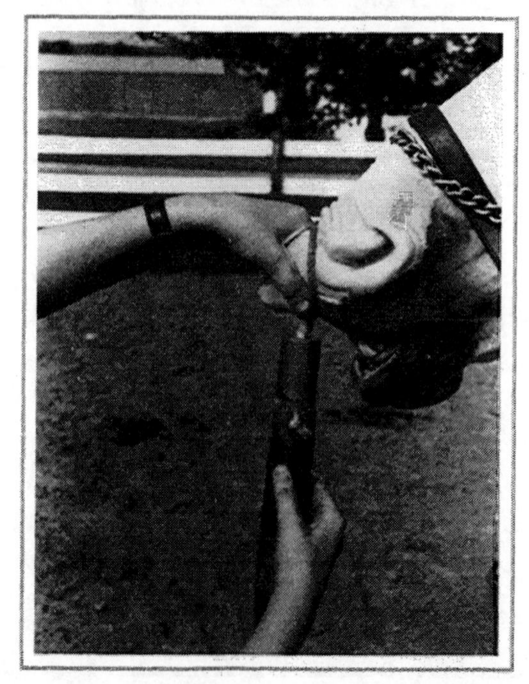

Figure 2-25 Rope encircling the upper lip

lip (see Figure 2-26). Care must be taken not to tighten the twitch too much, which can cause fear and avoidance reactions in the horse. If the handler has taken care not to get any fingers caught under the rope, he or she can now let go of the lip with the left hand and use it to control either the halter or the twitch handle (see Figure 2-27). After it is

Figure 2-26 Rope twitch in place

Figure 2-27 One version of controlling the twitch and the head

applied, it is helpful to make small vibrations with the twitch by making small rotations of the wrist of the controlling hand. For some reason this stimulation maximizes the effect of the twitch. It probably has something to do with the fact that a major acupressure and acupuncture point is located in the area where the twitch is applied. Like the twitch, acupuncture is thought to release endorphins. There may well be a connection.

Chain Twitch

The chain twitch (see Figure 2-28) is similar to the rope twitch except that the loop on the end of the handle is made out of chain instead of rope. Some people who work with horses feel that the openings in the links of the chain allow better circulation to the parts of the lip beyond the point where the twitch is applied. Although it could be argued that no twitches are particularly good for circulation, the fact remains that many people prefer the chain twitch. It is applied in the same manner as the rope twitch and is just as effective.

Kendal (Humane) Twitch

The Kendal, or humane, twitch is a device that pinches the upper lip of the horse by way of a hinge (see Figure 2-29). In this manner, the design prevents massive over-tightening, which can be a problem with novices using the rope or the chain twitch. The upper lip is grasped through the opening between the arms in a manner similar to the rope twitch (see Figure 2-30), and the arms pressed together, taking care not to

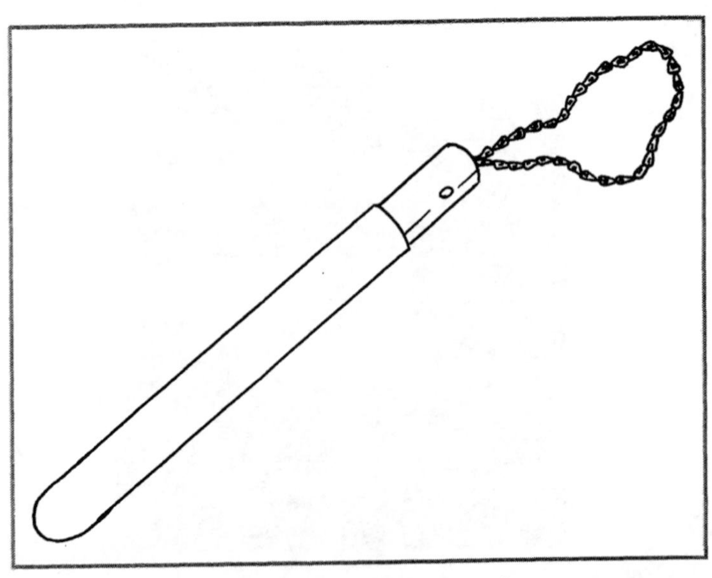

Figure 2-28 The chain twitch

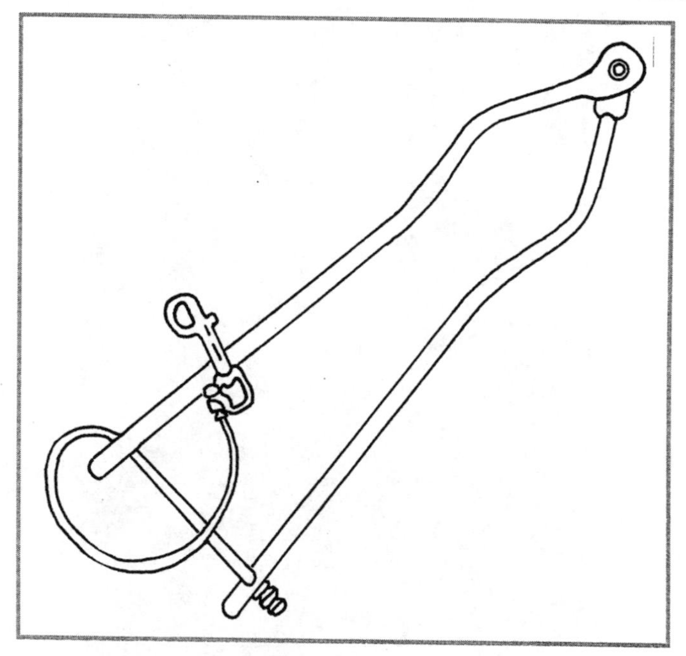

Figure 2-29 The Kendal, or humane, twitch

Figure 2-30 Preparing to apply the humane twitch

Figure 2-31 The humane twitch snapped in place

pinch the lip in the hinge. When provided with a string and snap, the humane twitch has some advantages. The string may be wound around the arms of the twitch, holding them closed, and then passed between the arms and snapped to the nearest halter ring (see Figure 2-31). It then holds itself in place, eliminating the need for a second person to control it. For someone who must work alone with the horse tied, this is a major advantage. Even when a handler is available, the humane twitch allows the handler to concentrate on other things, rather than the twitch itself.

Wilform Twitch

Another self-holding twitch is the Wilform. It is essentially a square piece of metal with a vice bar on one side that can be screwed in to pinch the upper lip when the square is fitted over it (see Figure 2-32). It is applied by reaching through the opening, grasping and holding the upper lip of the horse while turning in the screw (see Figure 2-33). When applied, it holds itself in place and has the advantage of no long arms for the horse to swing around and hit on things. As with all forms of twitch except the humane, care must be taken not to apply it too tightly.

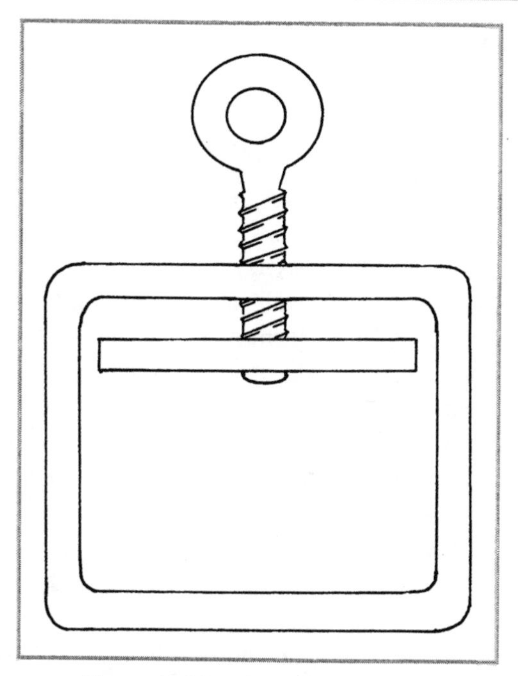

Figure 2-32 The Wilform twitch

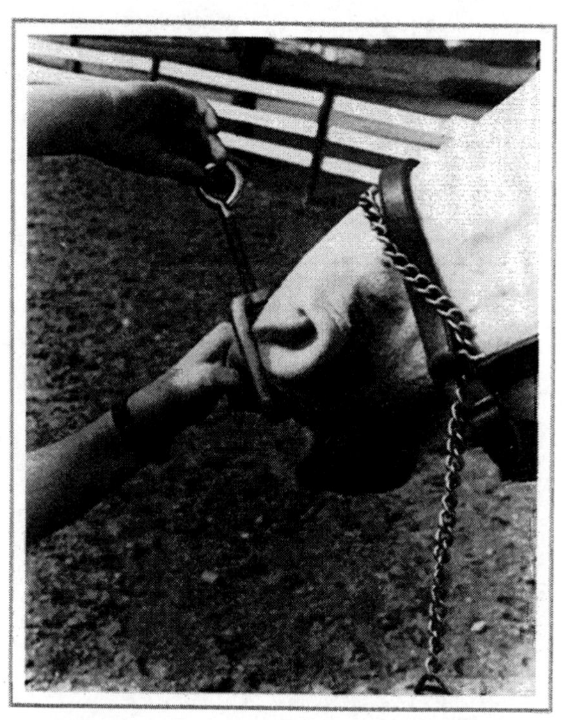

Figure 2-33 Applying the Wilform twitch

CHEMICAL RESTRAINTS

Chemical restraints in the form of tranquilizers are a major advantage in many situations. However, they should not be used by the untrained. They need to be administered in the correct amount for the body weight of the animal concerned. Consequently you should seek the aid of a competent veterinarian and follow his or her directions when administering them. It is possible for medical complications to arise after their application, which may take a veterinarian to handle. Some horses have negative reactions to certain tranquilizers, causing more trouble than you would have had if you had used a different form of restraint. Others do fine until they are coming out of sedation and then thrash around so badly that they injure themselves and others. Chemical restraints are useful tools, but they are not magic wands that solve all problems. If you need to administer them yourself, be certain to seek out quality instruction in the subject, and keep a good line of communication open with your veterinarian.

SUMMARY

As with any other tool used with horses, you should expect individual responses from animals to restraints. Some horses stand calmly and go to sleep when the twitch is applied; others panic and fight to the death in order to escape it. The same is true of the other restraints. Train your horses to accept them calmly before you need them in a real situation, and use good judgment regarding when and on which horses to use them. Around horses, there is no remedy for foolishness.

Catching, Leading, and Tieing

Approaching in an Open Space

Those who work with horses frequently have to catch loose horses, lead them to the area where they will be worked, and tie them up—provided the horse is trained to tie. Even if no work is to be done, the mere act of turning horses out for exercise and returning them to their stalls requires catching and leading. Because catching and leading are done so often, it pays to develop safe habits for doing so.

To catch a horse that is not actively resisting, handlers should first make sure that they have the halter and lead rope with them. I wish I had a nickel for every time someone (including myself) has gone out to the back fields to catch a horse only to find that the halter or lead rope was left back at the barn or that the halter was the wrong size. Some horses stand better if they don't see the halter and lead, so many people hold them behind their back with one hand as they approach.

As discussed in the first chapter, Zone B (next to the horse's shoulder) is usually the safest place to be around a horse, and this does not change when approaching one to catch it. Consequently it is best to approach the horse diagonally to the shoulder whenever possible (see Figure 3-1). This keeps the handler in Zone B and gives the horse a good look at the handler so that the horse isn't surprised. Surprising a horse usually frightens it, causing a defensive reaction that can get people hurt, so those experienced with horses develop ways of avoiding it. Some constantly whistle or hum

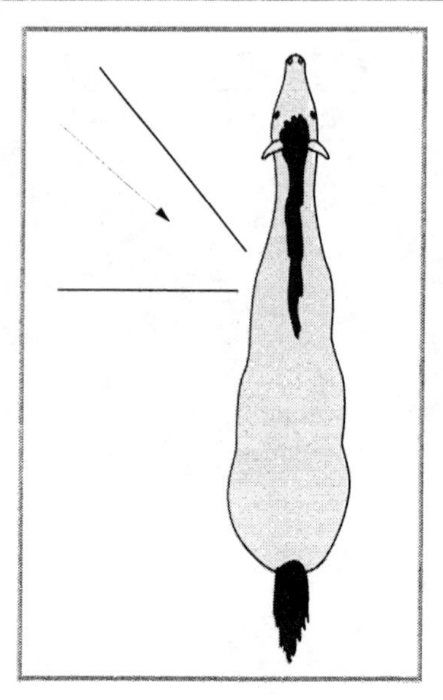

Figure 3-1 Approaching a horse diagonally to the shoulder keeps you in Zone B.

when they are around or approaching a horse, others mumble to themselves or actually talk to the horse. The tune they choose or what they say is not important within limits. What is important is that the horse knows exactly where they are at all times and that they are not acting in a suspicious manner. So it is usually best for the handler to speak to the horse first, and then approach through Zone B. The handler should avoid direct eye contact, facing the horse squarely, moving too quickly towards it, and grabbing quickly for the horse. These are all distance-increasing signals to horses that will make many of them move away. Instead, handlers should break direct eye contact, rotate their shoulders slightly, walk calmly and slowly (in a zigzag or serpentine manner for difficult horses), and avoid quick movements when they reach the horse (see Figure 3-2). This may not work every time, but at least they are not specifically telling the horse to move away. The trick to this entire procedure is that if the horse rotates or moves its body in any way, Zone B moves with it (see Figure 3-3). Approaching a horse properly often becomes a matter of jockeying for position so that the horse does not maneuver you out of Zone B. It is best to take your time and do it correctly, because staying in the correct zone limits the horse's ability to cause injury should it become frightened (see Chapter 1 for a refresher on safety zones). Resist the temptation to get sloppy because you know the horse, or because you have been working for years and nothing has happened to you *yet.* Every time you approach a horse, you are forming habits. As I once read, good habits last for years—bad habits last forever (or so it seems). It has been proven that when humans are under stress, they do what they have trained themselves to do. If you have trained yourself to always be precise and conduct yourself safely, you do so automatically even when you are so stressed that you can't consciously think of small details. So look at each approach to a horse as a chance to form habits that someday may save your life.

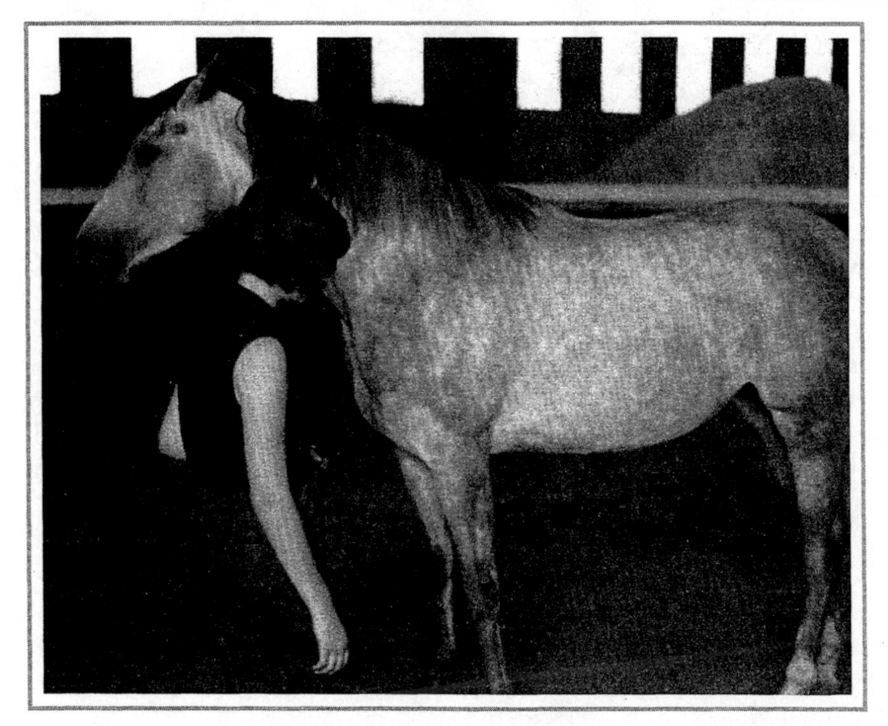

Figure 3-2 Approaching a horse

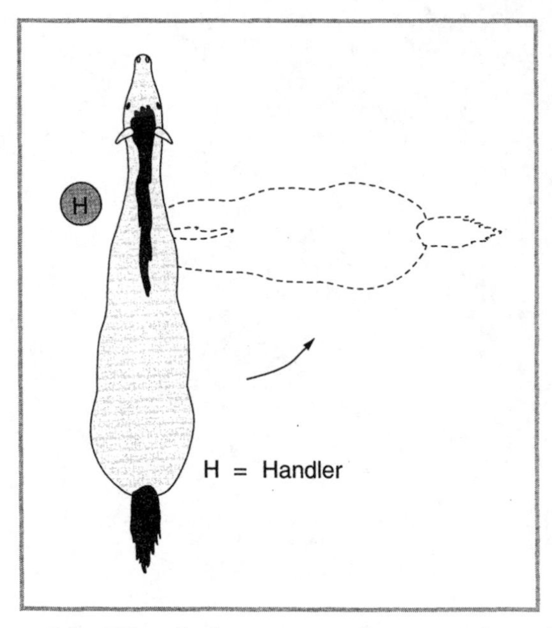

Figure 3-3 When the horse moves, the zones move with it.

Haltering in an Open Space

When the handler has successfully approached the horse, the first thing to do is to get the rope or something around its neck as close to the head as possible (see Figure 3-4). This discourages the horse from maneuvering, which makes it easier for the handler to stay in Zone B and, for some reason, convinces many horses that they are already caught. Why they think they are caught at this point is beyond me, but let's not argue with them.

If the horse does not have a halter on, this is a good time to halter it. It is usually best to place the noseband over the nose first for control before fitting the rest of the halter (see Figure 3-5). When the halter is in place, connect the lead to it, placing the chain, if there is one, in the appropriate position for the particular horse caught. By convention, this is usually done from the horse's left side (the near side), although it works from either side if done properly. If there is no halter, tie a bowline knot snugly around the horse's neck as close to the head as possible (see Figures 3-6 and 3-7). When tied correctly, the bowline knot will never slip. The rope itself will literally break before the knot will slip. Using this knot will eliminate the danger of choking the horse. All those who work with horses should practice this knot until they can tie it in pitch dark on a horse that won't stand still. Then put a half hitch over the horse's nose (see Figures 3-8 and 3-9), and the horse is ready to lead.

Figure 3-4 The lead rope around the neck

Figure 3-5 Placing the noseband over the nose

Figure 3-6 Illustration of the bowline knot around a horse's neck near the head

Figure 3-7 Bowline knot in place

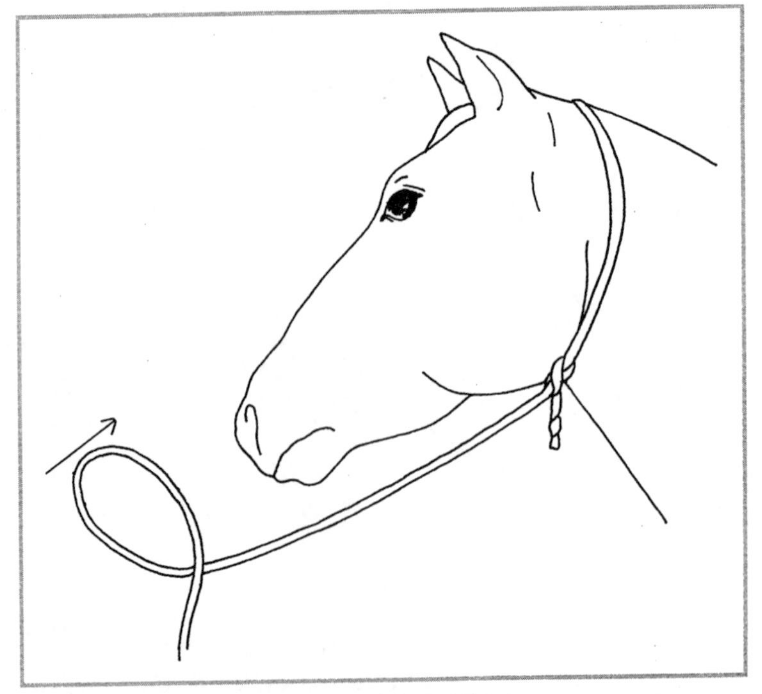

Figure 3-8 The half hitch

Figure 3-9 The half hitch in place

Catching Difficult Horses

Halter and "Handle"

Some horses are not this easy to approach. They may keep you at a distance when possible (knowing exactly how long your arms are), or even show aggression when you get too close. Some constantly compete to get you out of Zone B. These animals need to be handled and caught with care. For the borderline cases, it is useful to leave the halter on when possible, with a short length of rope tied to the lower halter ring (see Figure 3-10). This gives handlers something to grasp as soon as they reach the horse, and it convinces many horses that they will eventually be caught. They soon stop trying to evade.

Pole and Rope

Others are more determined to evade and are willing to hurt someone to accomplish it. For these horses, it is best to stay completely out of range of their legs and feet. A useful tool for this is a rope noose placed on and wrapped around a long pole (see Figure 3-11). If the pole is long enough, the noose can be slipped over the horse's head

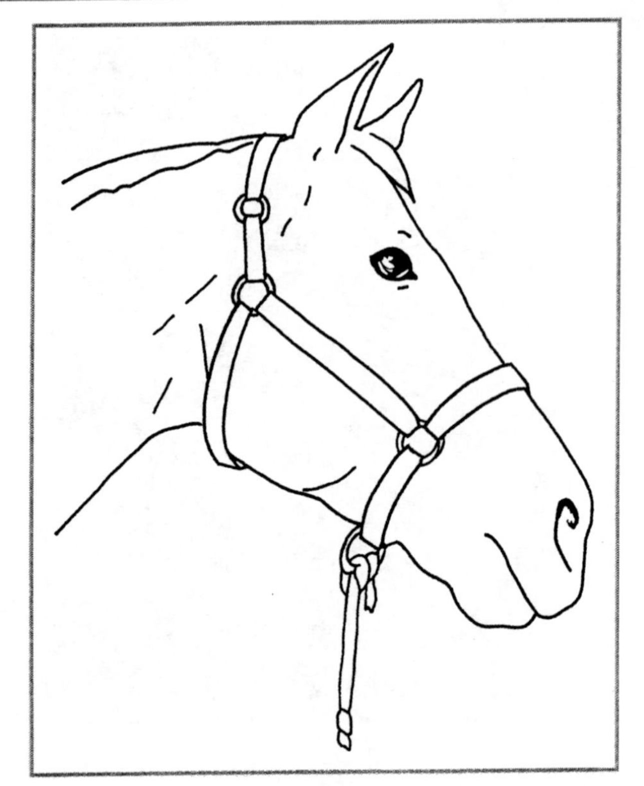

Figure 3-10 Short rope permanently tied to the halter ring

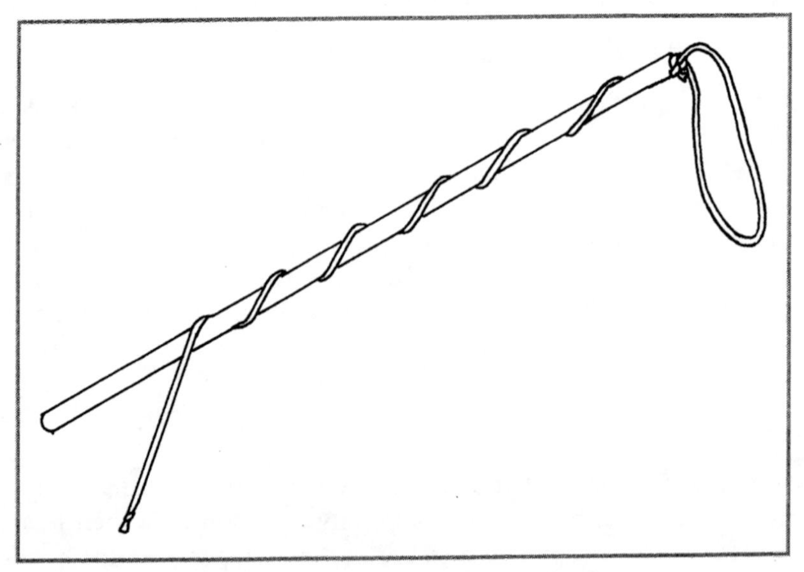

Figure 3-11 The rope noose and pole

even if it turns its rear end to the handler (see Figure 3-12). The pole can then be slipped off the rope and the horse's head pulled around to face the handler so that he or she can get to Zone B (see Figure 3-13), where the horse can be haltered and the lead shank connected in the appropriate manner. Horses that pull back frantically can cut off their air supply when you use a noose, so be ready to release the pressure on the neck before it becomes a problem.

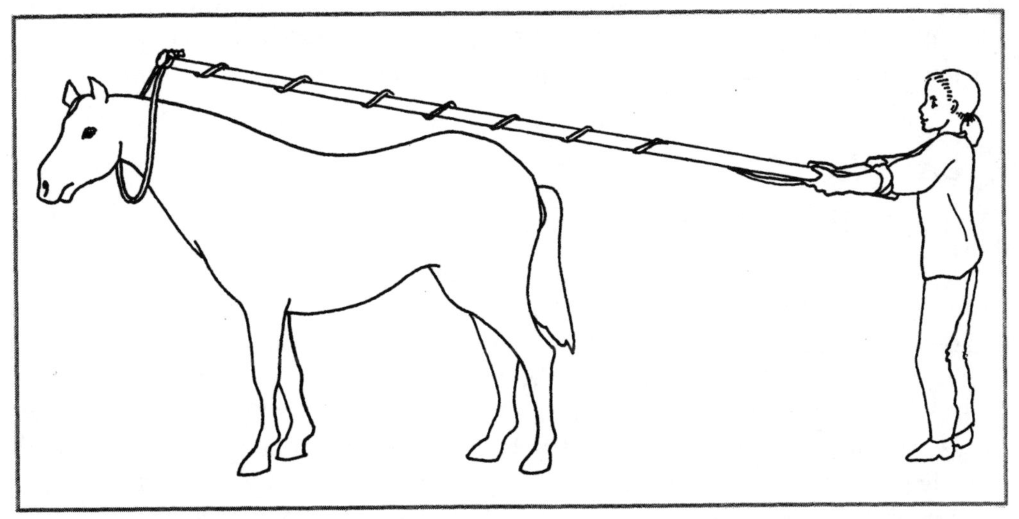

Figure 3-12 The rope noose and pole allow the handler to keep a safe distance.

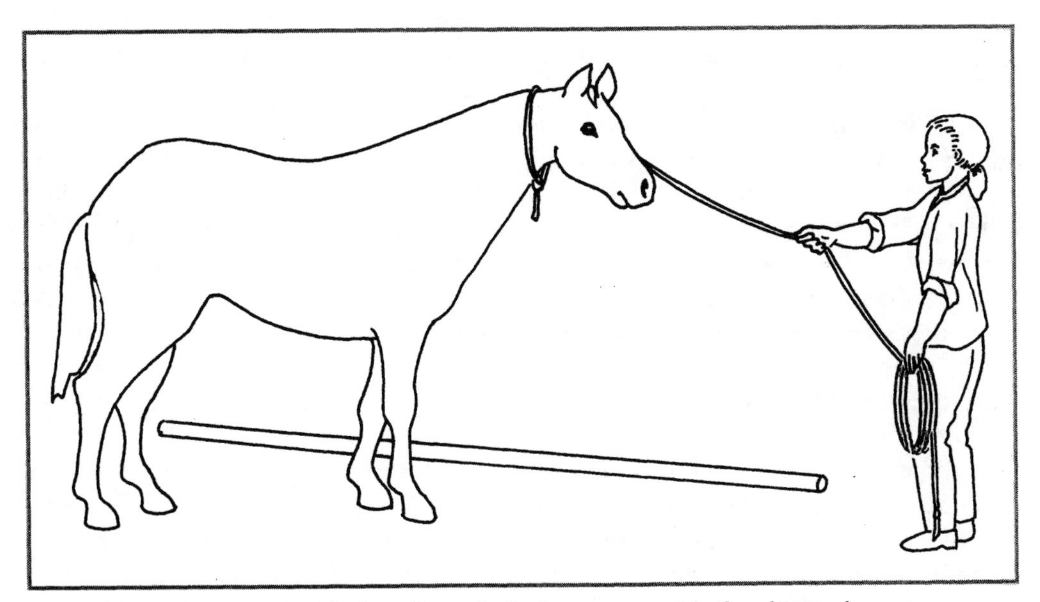

Figure 3-13 The handler pulls the horse around to face him or her.

Lariat

A faster method of using a rope is to become skillful with a lariat. Throwing a lariat encourages more movement from the horse, but this is an extremely fast way of catching difficult horses. If you want to keep the horse calm and stationary, however, learn to use the pole. In some cases, a tranquilizer dart gun may be used. This should be done with care, because the correct dosage is determined by knowing the horse's body weight and how much of the tranquilizer to administer per unit weight. Again, it is possible for some horses to have negative reactions to the chemicals, and it is easy for the untrained person to overdose a horse, so be sure to seek competent instruction before employing this technique.

LEADING

Using Halter and Lead

After the handler has caught the horse and haltered it, he or she is ready to lead it somewhere (unless it is tranquilized, in which case the handler may have to wait for the effects of the chemicals to wear off). When leading, it is important to remain in the safety zone (Zone B) as much as possible. It is also customary to lead from the near side (the horse's left side). So the handler should lead from the horse's left side, somewhere between the shoulder and the head when possible. When this is not possible, the handler should do what has to be done and return to the safety zone (Zone B) as soon as possible. If leading a horse through a very narrow opening, staying next to the shoulder can allow the horse to crush the handler against the frame of the opening (see Figure 3-14). In that case, it may be best to leave the shoulder and get in front of the horse momentarily to move through the opening first, returning to the shoulder position on the other side of the bottleneck (see Figure 3-15). There will be times when the handler will not want to lead from the left side. As a young man, I worked with horses in the backcountry of the Rocky Mountains in the western United States. While leading a string of horses across a difficult trail one day, I found myself leading from the left side on a narrow ledge with a drop-off of several hundred feet to my left. Being inexperienced in mountain horsemanship, I saw nothing wrong with this until my horse, whose head was to my right as it normally is, leaned over to rub its itchy head on me. This action instantly pushed me off the ledge, and only the reins, some bushes, and quick scrambling allowed me to extend my life span. The point here is that had I been sensible enough to lead from the off side (the horse's right side), nothing bad could have happened to me. It is good and sensible horsemanship to be able to lead and mount a horse from both sides.

Chapter 2 explains that it is safer to hold the lead about ten to twelve inches from the halter rather than the halter itself, because quick head movements can break fingers that are wrapped up in the halter. If the lead shank has a chain section, the handler should not hold the chain itself because this can also injure the hand if the horse pulls away quickly. The handler should not tie the excess lead to any part of his

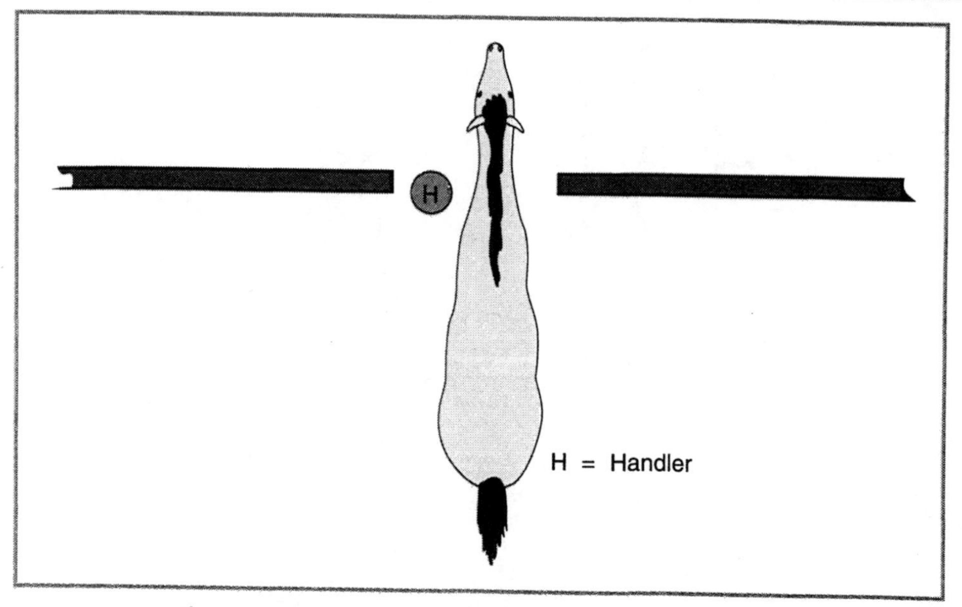

Figure 3-14 The handler at risk due to a narrow opening

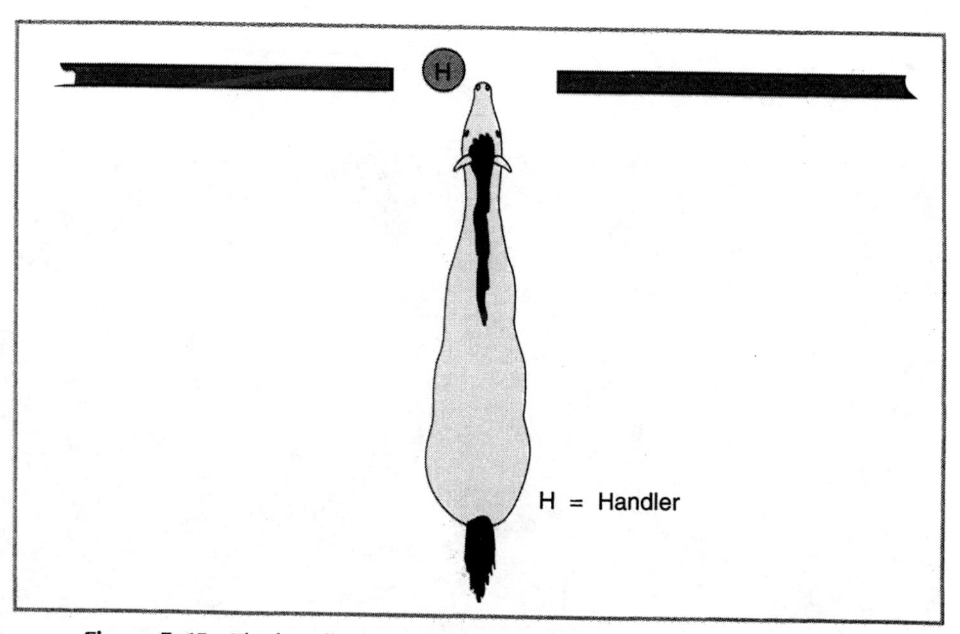

Figure 3-15 The handler leaving Zone B temporarily for a narrow opening

or her body or clothing or wrap it around a hand (the suicide grip). This allows the horse to drag the handler across the ground before he or she can get free. Being dragged can cause serious injuries. It is better to fold or coil the excess lead in the hand away from the horse (see Figures 3-16 and 3-17).

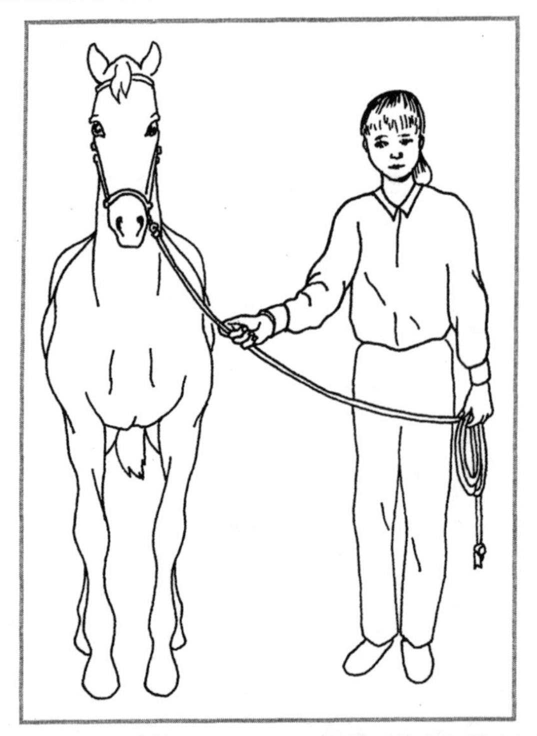

Figure 3-16 The excess lead held in coils in the outside hand

Figure 3-17 The excess lead held in folds in the outside hand

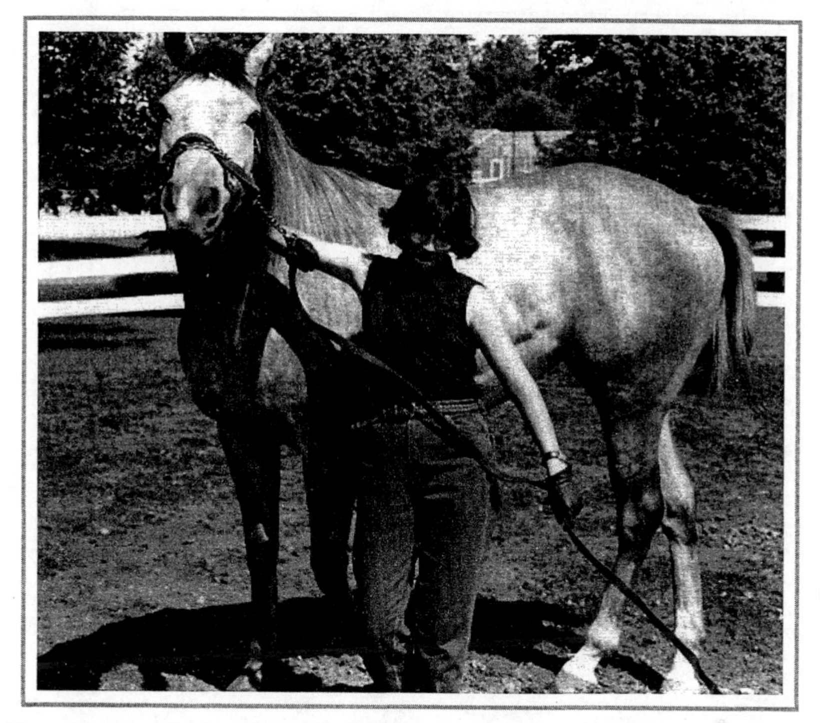

Figure 3-18 The handler extending the inside elbow during a sharp left turn

If the horse misbehaves while being lead, the handler should turn it sharply to the left if there is enough room to do so safely, and use whatever restraint is appropriate and necessary. Keeping the elbow nearer the horse extended helps keep the handler's feet from being stepped on during this procedure (see Figure 3-18). As mentioned in Chapter 1, turning to the left breaks the horse's forward momentum, bends the body, and allows the handler to stay in the safest zone (Zone B) until order is restored.

Passing While Leading

To keep other people safe, it is often best to lead the horse in such a way as to keep the handler between the horse and the people. This is because it is easier to turn the horse by pulling the head in the handler's direction than it is to try to push it away from the handler. So if handlers keep themselves between the horse and others, they can protect them from the horse's hind feet by pulling the head toward themselves, which will swing the rear end of the horse away from the people they are trying to protect (see Figure 3-19). If, on the other hand, the horse is between the handler and the other people, it can quickly swing its rear end toward them and kick, and there is relatively little the handler can do to protect them. Consequently, if you are leading from the near side of the horse, it is considered proper horsemanship to pass others by keeping them on your left side (see Figure 3-20) and inappropriate and dangerous to pass with them

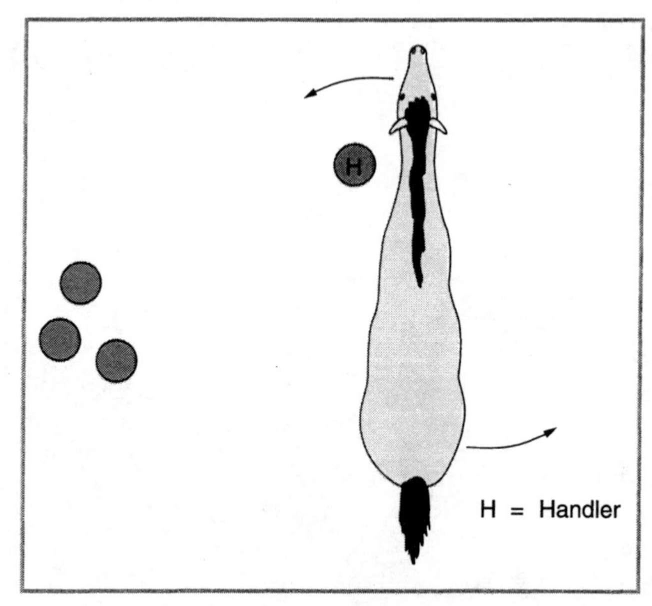

Figure 3-19 The handler protecting a group of people

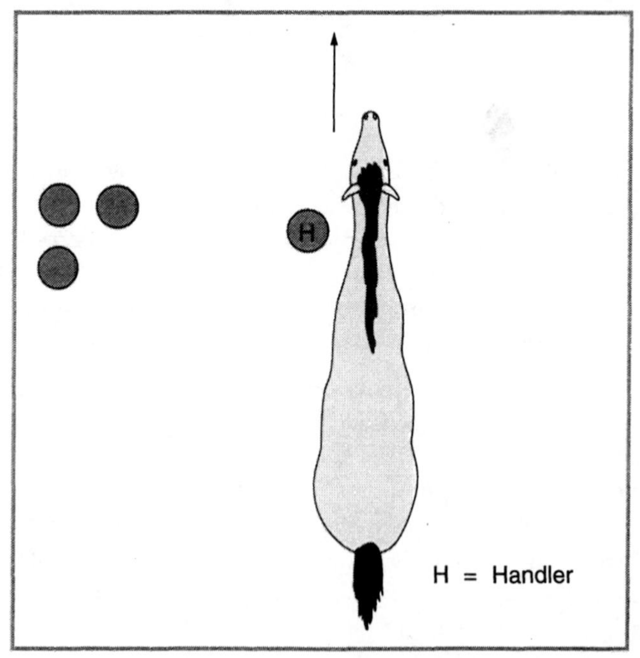

Figure 3-20 Proper positioning of the horse, handler, and a group of people

on your right side (see Figure 3-21). Obviously, if you are forced to lead from the off side (the horse's right side), you should pass others by keeping them on your right.

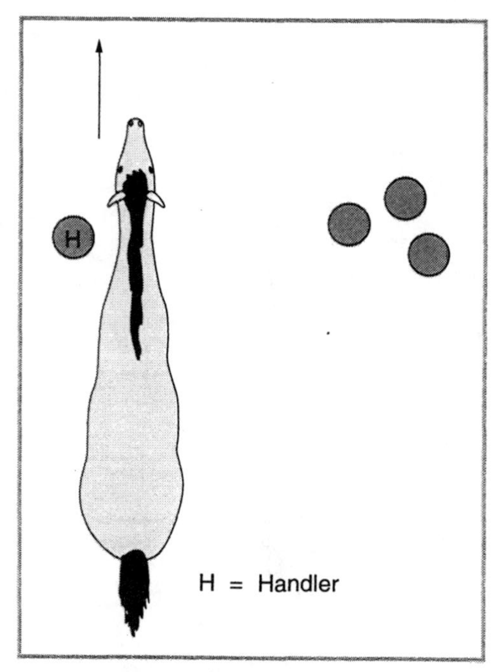

Figure 3-21 Incorrect positioning of the horse, handler, and a group of people

Backing Up on Lead

When asking a horse to walk backward, there are some items to remember. First, it is physically more difficult for a horse to move backward than it is for it to move forward. Therefore, many of them do so slowly, and some are quite reluctant to go backward at all. This tempts some handlers to step out of Zone B and position themselves in front of the horse to push it back (see Figure 3-22). This is not as safe as staying next to the shoulder (see Figure 3-23). The principles of steering remain the same when going backward, namely that the rear end goes the opposite direction from the head. So if the handler wishes to move the rear end farther to the horse's right, he or she must pull the head more to the horse's left (see Figure 3-24). It would be unusual for a handler to have to pass others while asking a horse to walk backward, but should the handler be forced to do so, he or she should be positioned on the same side of the horse as the people being passed. This will allow the handler to swing the rear end away from the others by pulling the head toward himself or herself. However, it would be much better to turn the horse around and lead it past them in the forward direction if this were possible.

Horses That Rear

If a horse has a tendency to rear up with the front end while being lead, it is wise to use a longer lead than normal. A normal length of lead can be pulled out of the handler's hands when the horse goes up in the air. It is also difficult to stay in Zone B

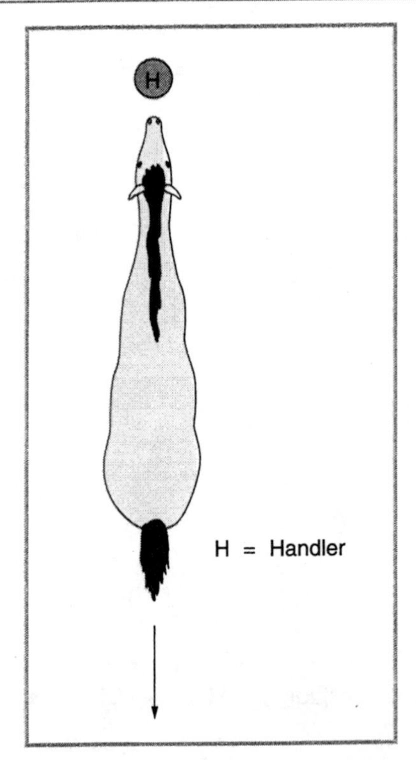

Figure 3-22 Incorrect positioning of handler to move a horse backward

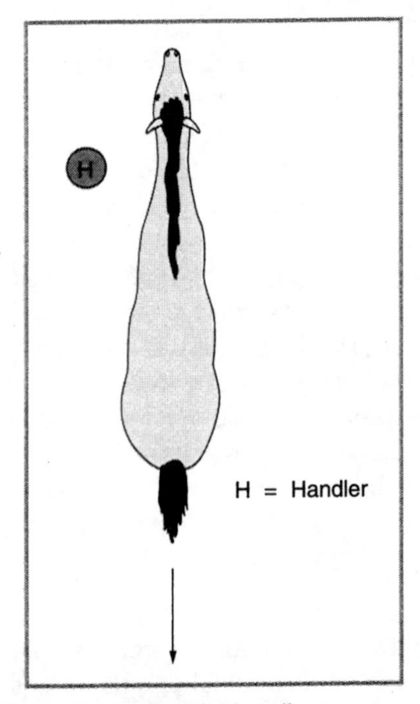

Figure 3-23 Safer positioning of the handler to move a horse backward

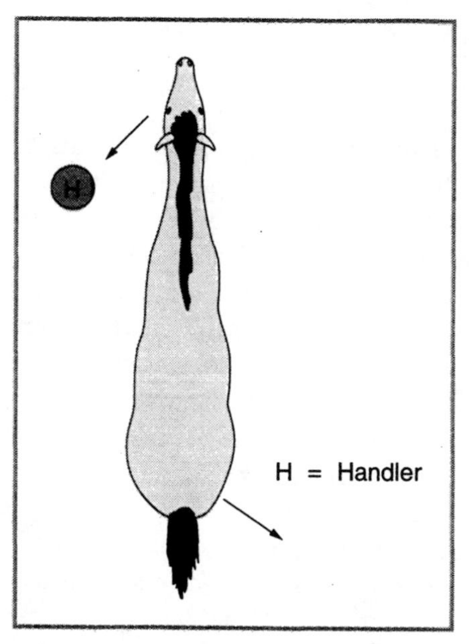

Figure 3-24 Steering to the horse's right when moving backward

under these circumstances, and sometimes the best thing to do is to use the longer length of lead to back away completely out of range of the front feet until the danger is past (see Figure 3-25).

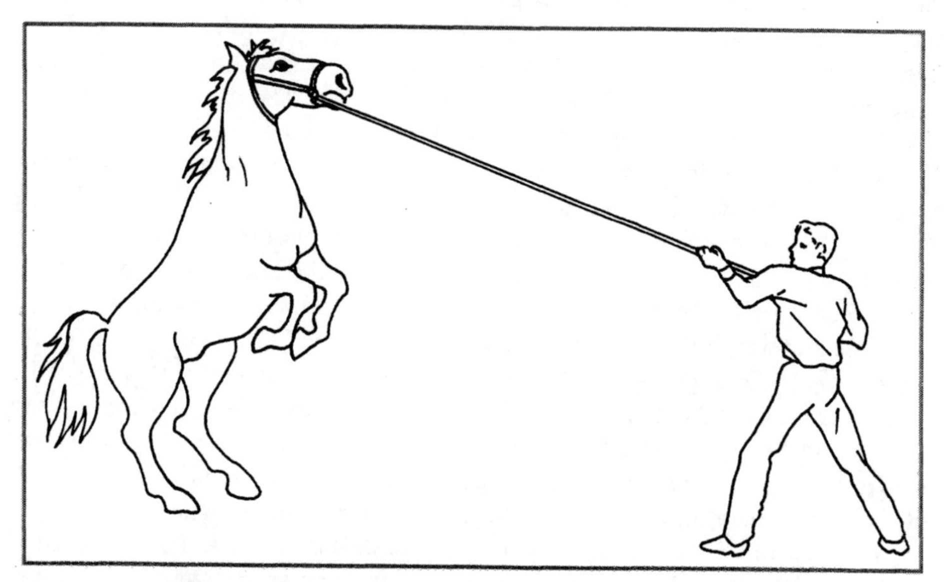

Figure 3-25 Handler backing away from a rearing horse

TIEING

To Tie or Not to Tie

When the handler reaches the work area, it is an advantage to be able to tie the horse to something so that the handler can concentrate on the work at hand rather than hold the horse. Sometimes the handler just wants the horse to remain in one spot until he or she returns. Whatever the reason, tieing up a horse can either be a pleasure or a nightmare, depending upon how the horse reacts to it. Horses that are frightened of the sensation of being tied solidly can panic, throwing themselves backward, falling over on their sides, causing severe harm to themselves, and generally trampling any humans who get in their way or try to stop them. Handlers should not tie strange horses to anything until they know whether they "tie." If there is no one around who knows the horse, the handler should experiment first, by tieing the horse and remaining with it, ready to release it instantly should problems occur. Only after it has proven itself should a horse be tied and left for any amount of time. Lastly, young horses should be taught to stand calmly when tied, and no horse should be tied by the reins that are connected to the bit in its mouth, regardless of what you see on TV or in the movies. Reins usually break under pressure, but not before they injure the mouth.

Single Tieing Point

If the horse is to be tied to a single tieing point, the handler should prepare the horse by removing any halter restraints that are in place. Do not tie a horse with a chain over the nose, under the chin, in the mouth, or over the gums (see Chapter 2 for a review of these restraints). If the horse were to move backward, the chain would create pain when the lead tightened. This might panic the horse, and if it threw itself backward against the restraint, it could injure itself. The only connection to the horse should be the halter with a lead connected to the halter ring under the chin. When the lead and halter are in the proper condition, the handler must select an appropriate object to tie the horse to. In a barn situation, there are usually tie rings placed in optimum locations. If the handler is outside the barn environment, he or she must choose with more care. For instance, dead branches and trees or rotting posts and crossrails do not make good tie points. It is no fun watching helplessly as the horse breaks the post and bolts frantically away with a rotted four-foot section tied to the end of its lead. The more it runs, the more noise the post makes and the more it bangs into the horse's body. The more noise and banging occurs, the faster the horse runs. I think you get the picture—so choose well. Tie to green, live trees and branches of sufficient size, and good solid posts and crossrails. The tieing point should be at least as high as the horse's withers (the point on the top of the horse where the neck and back meet). This is to keep the horse from stepping over the lead after it is tied. When a horse is tied too low, it can easily step over the lead. Its legs then become entangled, and it usually frightens itself enough to rear and thrash around, causing a good deal of damage. The handler should use a good quick release knot (see Figure 3-26). When an animal weighing a thousand

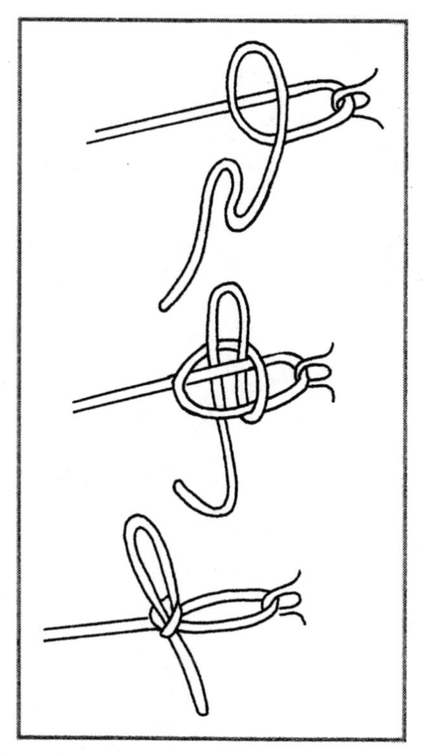

Figure 3-26 Quick release knots should be used when tieing horses to a solid object.

pounds or more throws itself back on the lead, some knots do not release easily. Many knots that seem functional with no stress do not release at all under the level of pressure a horse puts on them. The handler should leave only enough lead between the horse's head and the tieing point for the horse's head to reach the ground—no more. Many behavioral experts feel that it is important for horses to be able to lie down, because they only get Rapid Eye Movement (REM) sleep when they are in that position, but any excess lead gives them more chance to step over the lead with a front foot. If there is no halter available, the handler should tie a bowline knot around the horse's neck snugly as close to the head as possible (see Figure 3-6) but not touch the horse's nose with the lead. He should then tie normally (see Figure 3-27).

Crossties

Another way to tie a horse is to use what are called crossties. Instead of one line tieing the horse, there are two lines, one from each side of the head connecting to the halter rings (see Figure 3-28). Keep in mind that some horses tie nicely with one line but have never been put in crossties. Some will panic and hurt themselves, even though they stand well for a single tie. So monitor horses carefully the first time you put them in crossties. This is best done by hooking up one line and leaving your lead line on the other side of the halter. You can then imitate the pressure that the second line will

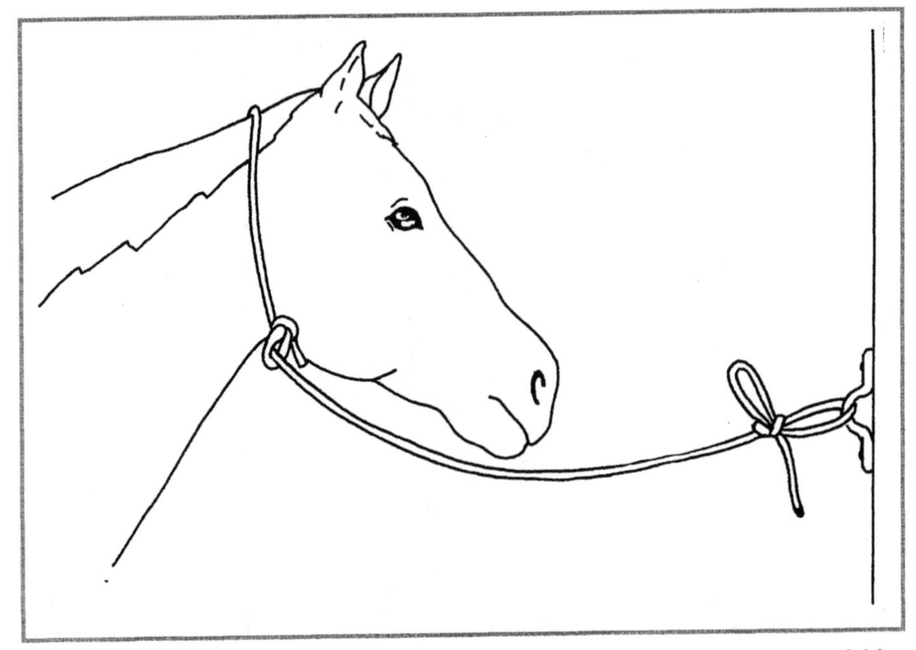

Figure 3-27 Tieing with a bowline and a quick release when no halter is available

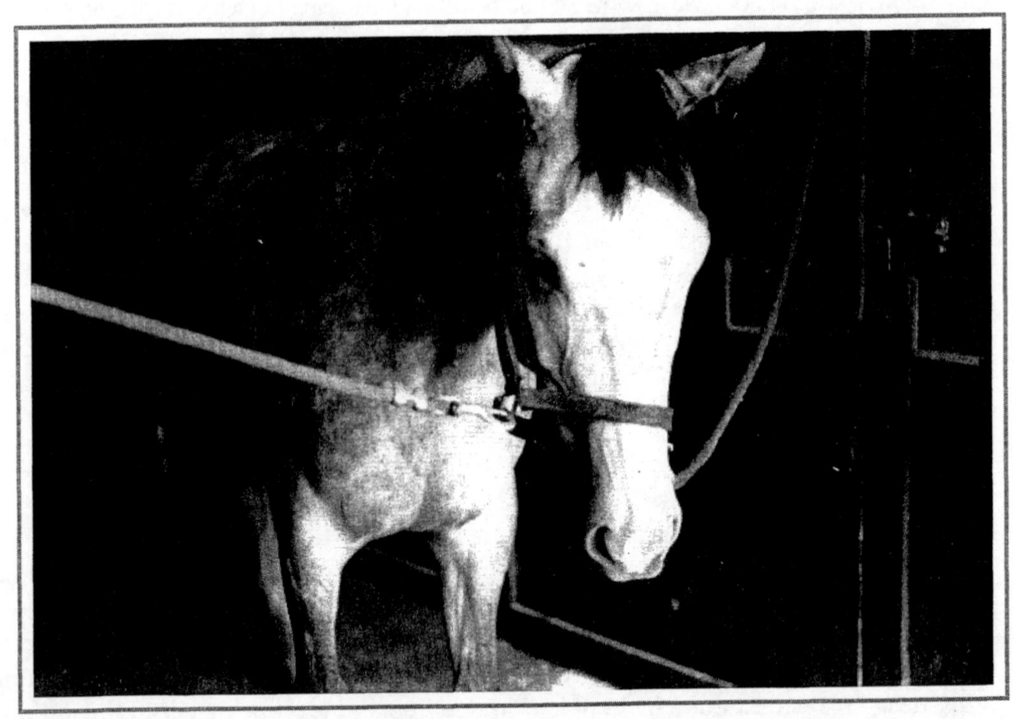

Figure 3-28 A horse in crossties

cause and if the horse reacts negatively, you can release it immediately. Another approach is to use bailing twine or some easily breakable material for the crossties so that if the horse panics, it can easily break free without hurting itself. After the horse has proven itself, crosstieing is simple. The handler hooks up one side of the halter and removes all restraints and then reads the horse quickly to see if everything is normal. If it is, the handler hooks up the second line, reads the horse again quickly, and goes about regular business. The lines may be tied into the side halter rings using quick release knots (see Figure 3-26) or connected using snaps. If snaps are used, it is best to have special quick release snaps for the same reason that quick release knots are used.

Summary

In many living environments, horses are constantly being turned out, caught, and brought back in. They spend a considerable amount of time tied while they are groomed, shod, saddled, harnessed, or worked on by the veterinarian. The time you spend training your horses to stand calmly in ties is well invested. The time you spend teaching your human employees or helpers to handle them safely is equally well spent. The horses are better cared for, and the humans are safer. This increases efficiency and saves money in the long run. I think that's called good management.

CHAPTER 4

Stalls and Paddocks

Box Stalls

When being kept inside a building, horses are often housed in spaces that are wide enough for the animal to turn and move around in. There is usually a door or barrier of some sort at the entranceway, and often the horse is left loose with the door shut. Because the dimensions are often square or rectangular in nature (resembling a box), these enclosures are commonly referred to as box stalls (see Figure 4-1). Like anything else, they have advantages and disadvantages. With regard to safety, the advantage is having more room to work around the horse. This lowers the chances of you being accidentally caught against the wall of the stall. The disadvantage is that if you ever get isolated in one of the back corners of a box stall with the horse between you and the door, you are completely cut off and at the mercy of the horse. Fortunately this is not a frequent problem and can be eliminated by using proper procedures.

To get a horse out of a box stall, first be certain that the horse knows that you are coming. Many animals react defensively when surprised at close range, which is not usually good for human safety. Speak to it if possible, and pay attention to its body language to see if it is calm or agitated. If the situation looks safe, open the door slightly and try to get the horse to face the door by speaking to it or making some other interesting noises. If it insists on keeping its rear end towards you, and lifts its hind feet occasionally to threaten you with a kick, stop. You want the horse to face you before approaching it, because you want to approach through Zone B, not Zone D. If the rear end is all the horse presents to you, you cannot approach safely. So wait until the horse faces you, and train your horses to do this if necessary. Open the door just

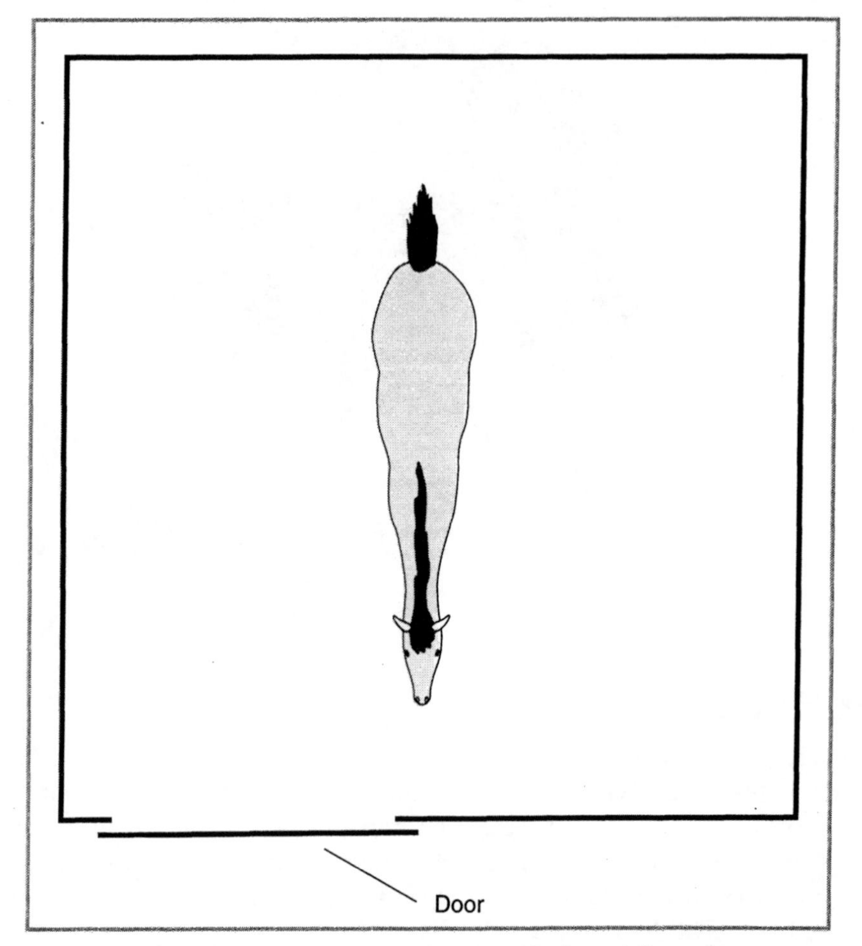

Figure 4-1 A representation of a box stall

wide enough for you to enter, but not wide enough for the horse to exit. Enter the stall in such a way that you remain between the horse and the door as much as possible. It is safer for you to have access to the door in case you need to exit sooner than you expected. This way if the horse becomes frightened or starts behaving badly for some other reason, you can simply take a step or two back and be out the door. For this reason, do not close the door behind you, but leave a small opening in the doorway (large enough for you to squeeze through, but not wide enough for the horse to think it can pass through). If the horse is not haltered, halter it and connect the lead shank in the appropriate manner. Announce to everyone in the aisle that you are exiting the stall. This is to ensure that no one is caught by surprise and that someone can warn you not to come out if there is a problem. Open the door completely, checking to make sure that the latch is not sticking out where the horse might scrape its side against it (see Figure 4-2), and exit the stall, watching the horse's hips to see that they do not bang against the door or the door frame as you do so.

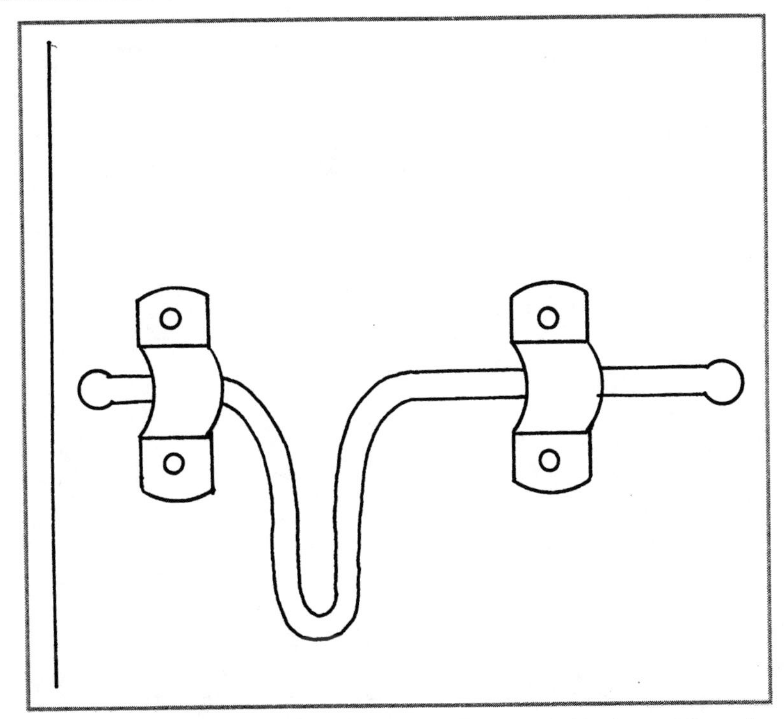

Figure 4-2 Make sure that the door latch is fully retracted before exiting a box stall.

When it is time to return the horse to the box stall, begin by making sure that the door is completely open. Banging against the door on the way in will frighten most horses, which can create problems. Lead the horse in carefully and turn it around so that it is facing the door. As discussed in Chapter 1, the direction you turn a horse is important. If you are leading the horse from the horse's left side, the near side, it is best to turn the horse to the left, to keep yourself on the inside of the turn. This keeps you in Zone B and also keeps you from getting cut off from the door (see Figure 4-3). Naturally, if you are leading the horse from its right side, the off side, you would do the opposite, turning to the right. When you have completed the turn and returned to the door with the horse facing it, close the door partially, but not completely (see Figure 4-4). You can then release the horse by removing the lead shank and halter. Do not release the horse until you have it facing the door, because once again you want to have access to the door without the chance of being cut off and trapped in one of the back corners. You also do not want to have to pass the rear end of the horse to reach the door, because that would take you through Zone D. As soon as the horse is released, step back through the door and close it.

If it is necessary for you to work on a horse in a box stall, it is usually best to restrict its maneuverability. You could have someone else hold it, put it in crossties, or tie it to a single tieing point. If the horse is restricted, the handlers always know where Zone B is in case they can't reach the door.

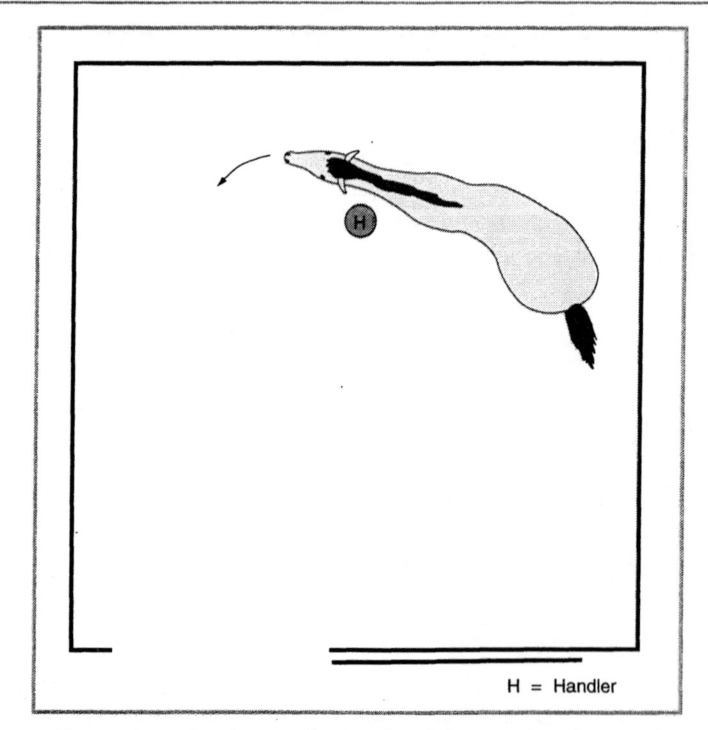

Figure 4-3 Staying on the inside of the turn in a box stall

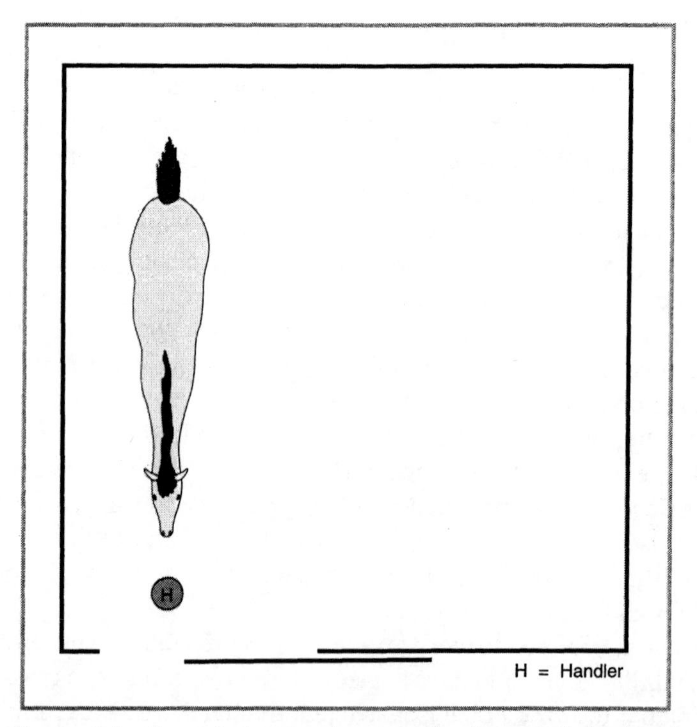

Figure 4-4 Horse facing the door and the door partially closed

STRAIGHT OR TIE STALLS

In contrast to the box stall, the straight stall (often referred to as the tie stall) has smaller dimensions. It is often little more than two horses' widths across, and not much longer than the length of a large horse (see Figure 4-5). It is frequently equipped with a chain across the entranceway, called a butt chain. It appears much straighter in shape than the box stall, and the horse is often tied to a single tieing point at the front, hence the names "straight" or "tie" stall. Like the box stall, the straight stall has advantages and disadvantages. In terms of safety, it is easier to get to the exit in a straight stall and there are fewer maneuvers to worry about, but there is the disadvantage of less room, making it easier to get pinned up against the wall. It also requires you to approach the horse through Zone D, because the head is usually tied at the other end of the stall.

To get a horse out of a straight stall, approach slowly and speak to the horse so that there are no surprises. Standing behind one of the side walls, gently push the rear end of the horse over to make room at the entranceway (see Figure 4-6). If the horse refuses to move over, do not enter the stall. Wait until you can think of some way to get the horse to move over, and if this uses up too much time, start training the horse specifically to perform this maneuver. Remember, you are going to move through Zone D, so you want everything to be working smoothly before you enter. After the horse has moved over, stand up straight as you remove the butt chains. Do not lean down, because this will put your face and head too close to the horse's hind feet. Walk up to the horse's head, keeping a hand on the horse's side to remind it not to move in your direction. When you reach the head, connect the lead shank in the appropriate manner

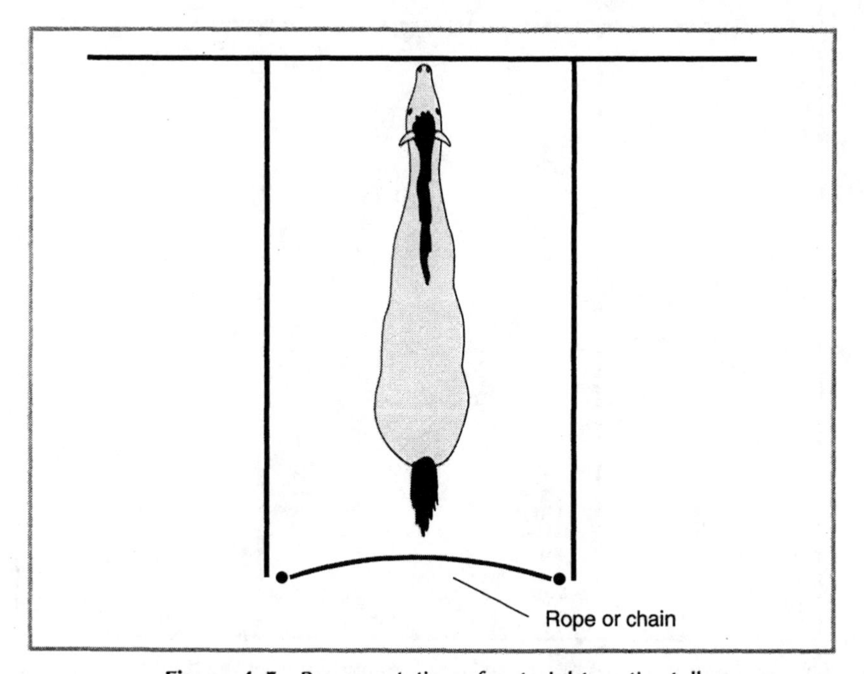

Figure 4-5 Representation of a straight, or tie stall

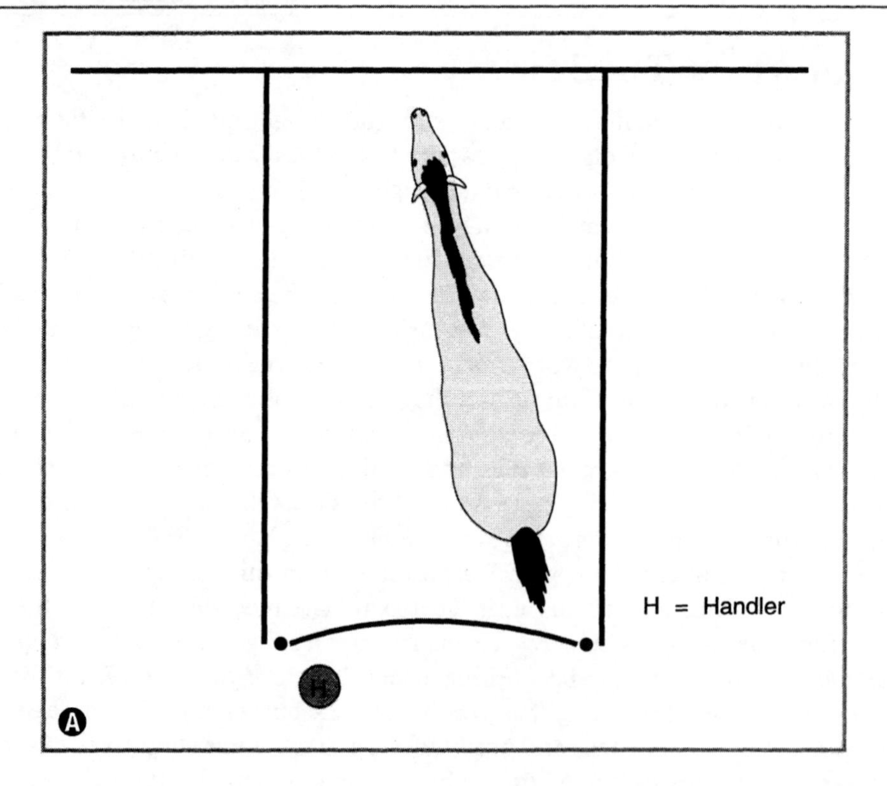

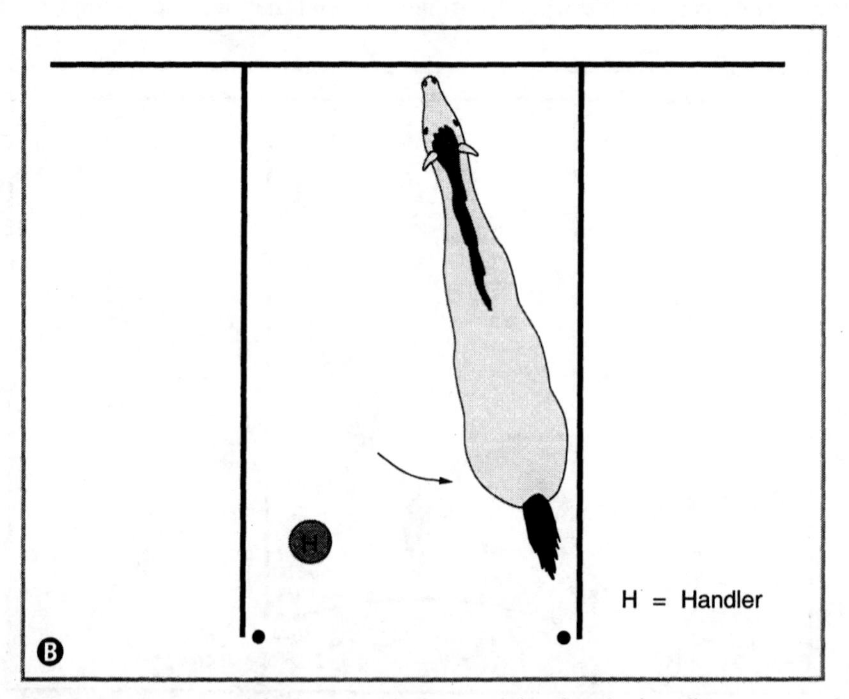

Figure 4-6 Proper position of the horse (a) before and (b) after the handler enters straight stall

before untieing the horse's head. Announce your intention to exit the stall and check visually, in case someone did not hear you. If no one objects and you see no problems, back the horse slowly out of the stall (see Figures 4-6, 4-7, 4-8, and 4-9).

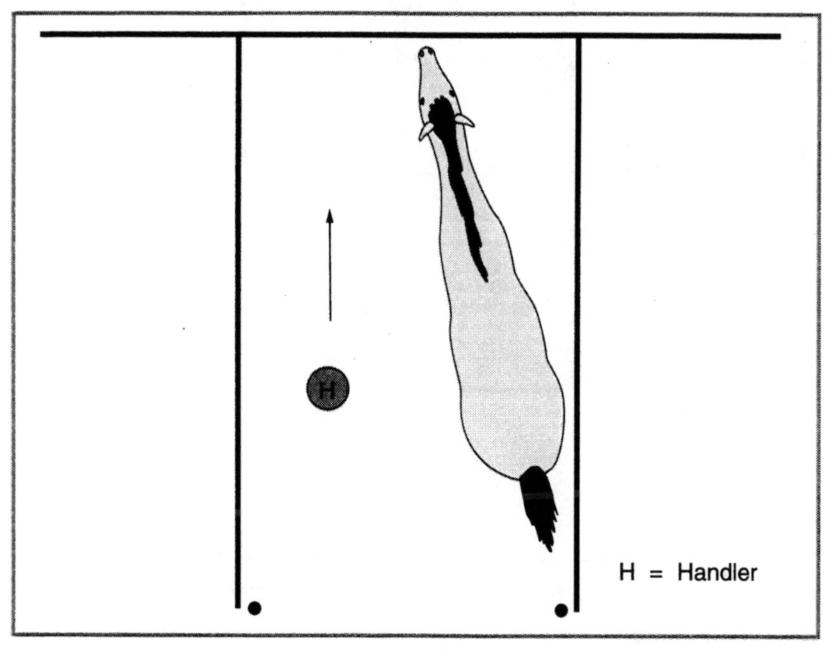

Figure 4-7 Approaching the horse's head in a straight stall

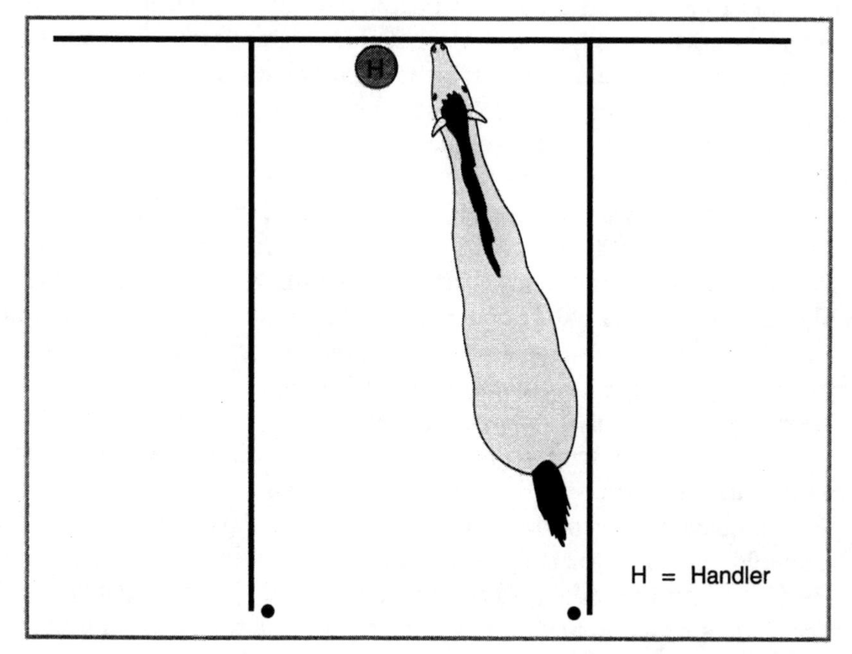

Figure 4-8 Handler's position when connecting the lead shank and untieing the horse

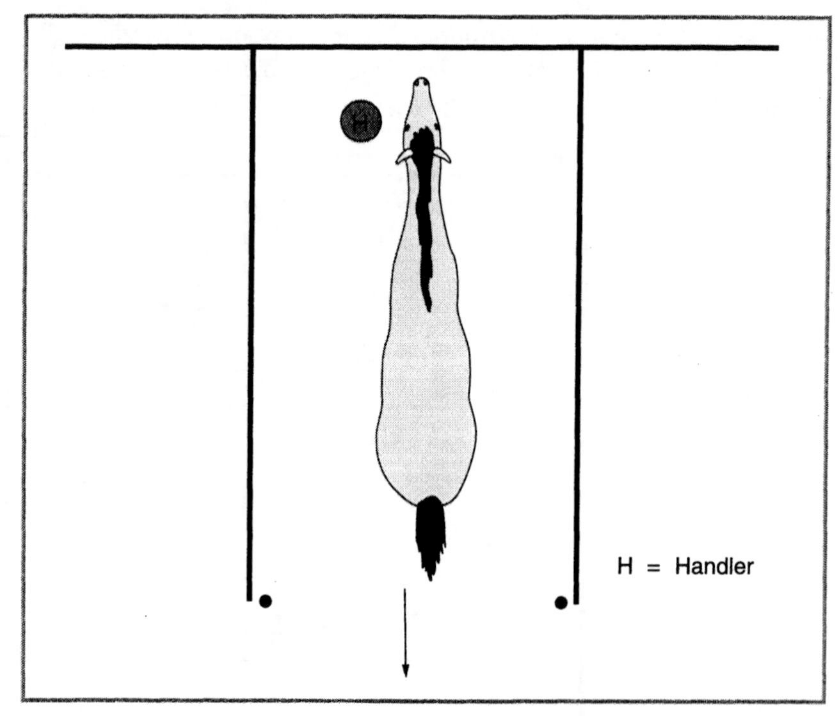

Figure 4-9 Backing out of a straight stall

Some horses are uncomfortable in straight stalls unless their heads are left untied. To remove an untied horse from a straight stall, announce your intentions and check for problems in the aisle before you remove the butt chains. Then step aside and allow the horse to back out of the stall itself, catching the halter as it reaches your position at the end of the stall. If the horse does not back out itself, push the rump over, approach the head, and remove the horse normally.

To return a horse to a straight stall, make sure that the butt chains are out of the way and lead the horse straight into the stall and up to the tieing point at the front. Tie the head in the appropriate manner before removing the lead shank. Exit the stall smoothly, keeping a hand on the horse to remind it not to move in your direction, and replace the butt chains (see Figures 4-10, 4-11, and 4-12). If the horse stands loose, follow the same procedure, eliminating the head tieing.

Working on a horse in a straight stall should only be done with calm, reliable horses. If there is any question regarding the horse's suitability, it is best to remove it from the stall and work on it in the aisle or some other location that offers more space. Horses being worked on in a straight stall should have their heads tied, and the handler should remember to change sides of the horse by exiting the stall, pushing the horse over, and reentering the stall on the opposite side. Switching sides by ducking under the horse's neck, tie line, or belly often frightens the horse, causing a dangerous situation.

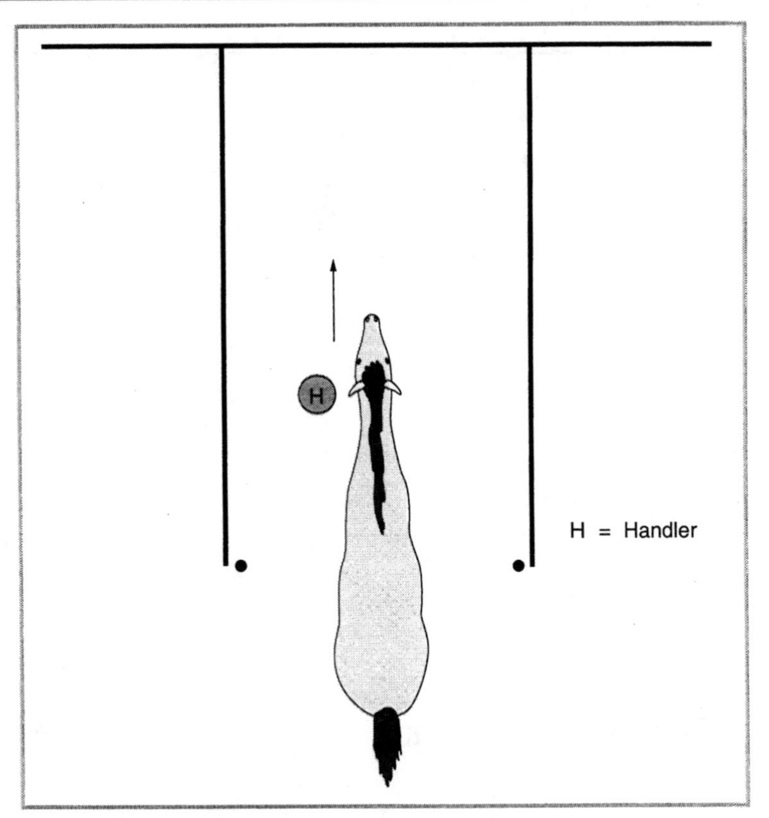

Figure 4-10 Entering a straight stall

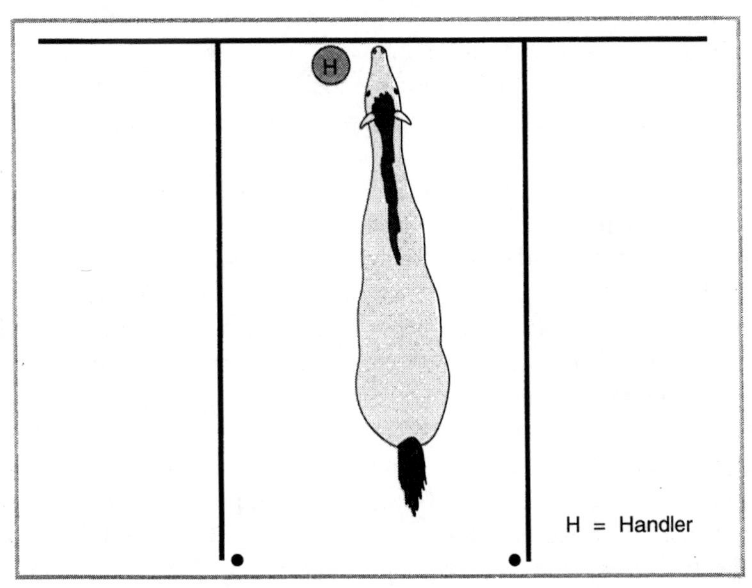

Figure 4-11 Position of handler and horse for tieing up and removal of the lead

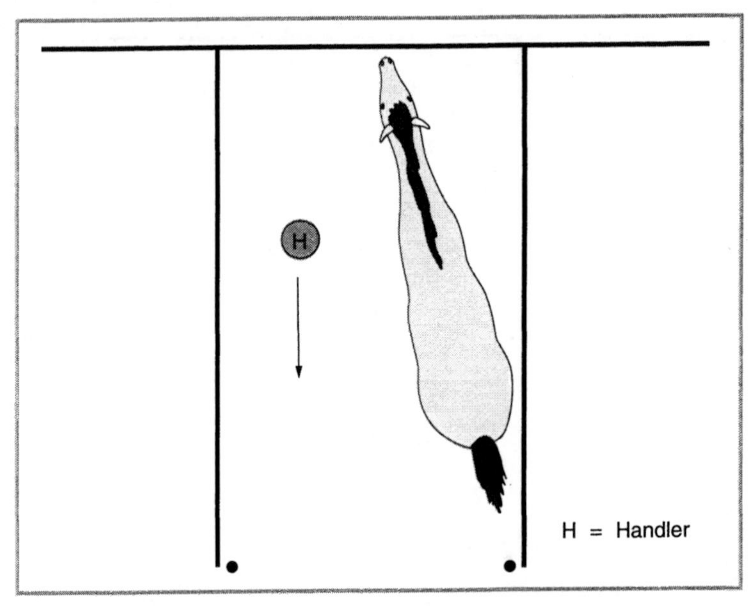

Figure 4-12 The handler exiting a straight stall

PASTURES AND PADDOCKS

Pastures are fenced, outdoor areas that often (but not always) support some amount of forage for the horse to graze on while it is enclosed there. Paddocks, on the other hand, are similar except that there is no expectation of forage for the horse and they are often thought of as being smaller in area than pastures. They are associated with shorter stays, such as a daily turn out for exercise and other short-term activities. Both areas have enclosing "walls" (fences, actually) and one or more "doors," which we usually refer to as gates. As such, they are similar to box stalls in design, and safety procedures will reflect this. Indeed, many people think of them merely as large box stalls and act accordingly.

Before removing a horse from a paddock or pasture, you may wish to review safe catching and leading procedures as described in Chapter 3. After you have done that, adapt them to the number and type of horses you are dealing with. For instance, if there are multiple horses in the paddock, it may be safest to have some helpers with you. Then there are people to catch the horses that are to be removed, and still others to occupy the attention of the horses to be left behind. This reduces the temptation some horses have to play with or heckle the ones that have been caught. It often appears that the hecklers mean no harm, but you still get stepped on or knocked down. If you must work alone, you may wish to carry a crop or short whip to discourage troublemakers.

If the horse you need is alone and cooperative and walks over to the gate, you can often catch it merely by opening the gate a small amount, stepping in, and connecting the lead rope. If it is not alone or not cooperative, the procedure may take longer, so it is customary in that case to enter the paddock and close the gate behind you, unless you have helpers to control the gate. After you have caught the horse and connected the lead rope, you return to the gate, reopen it, and lead the horse through. Controlling the gate as well as possible, turn the horse around to face it while you close it again. You then turn and proceed to your destination (see Figures 4-13, 4-14, and 4-15).

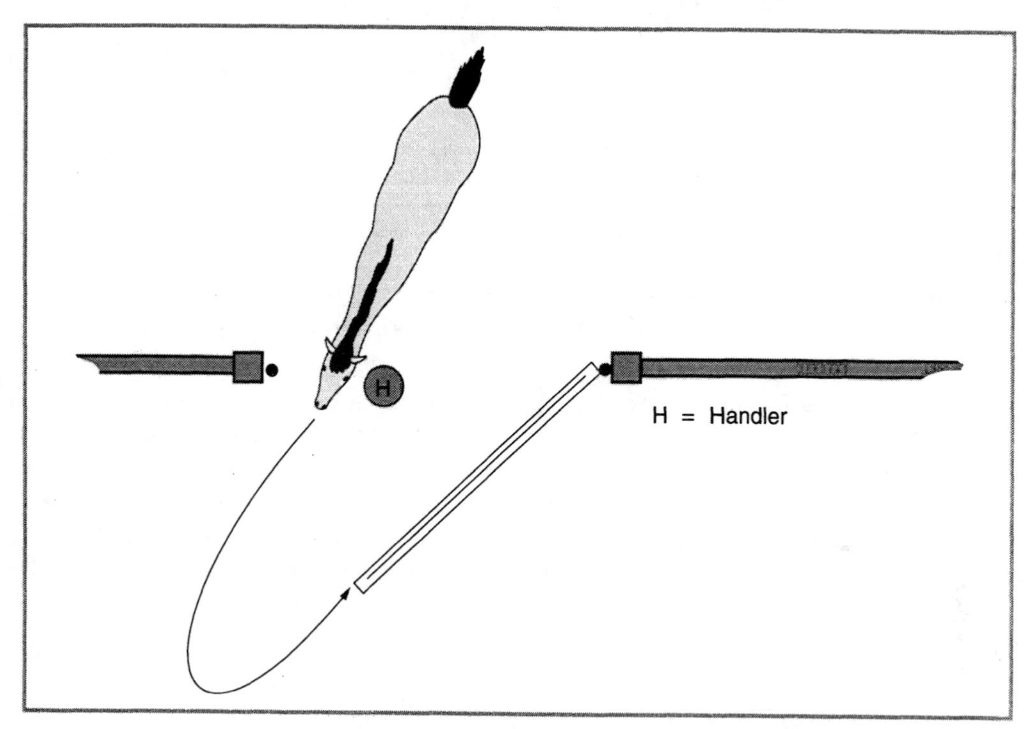

Figure 4-13 Leading through the gate and turning to face it

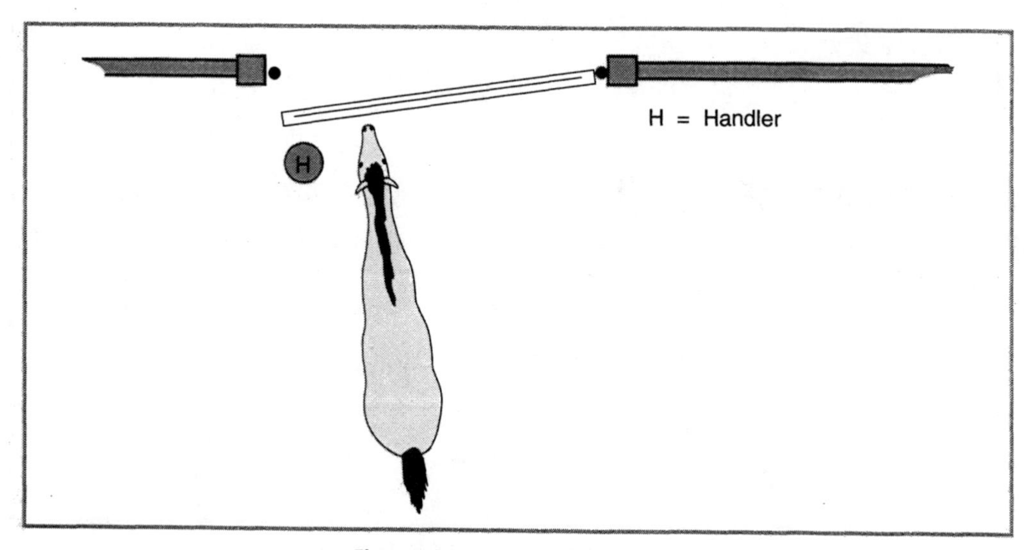

Figure 4-14 Reclosing the gate

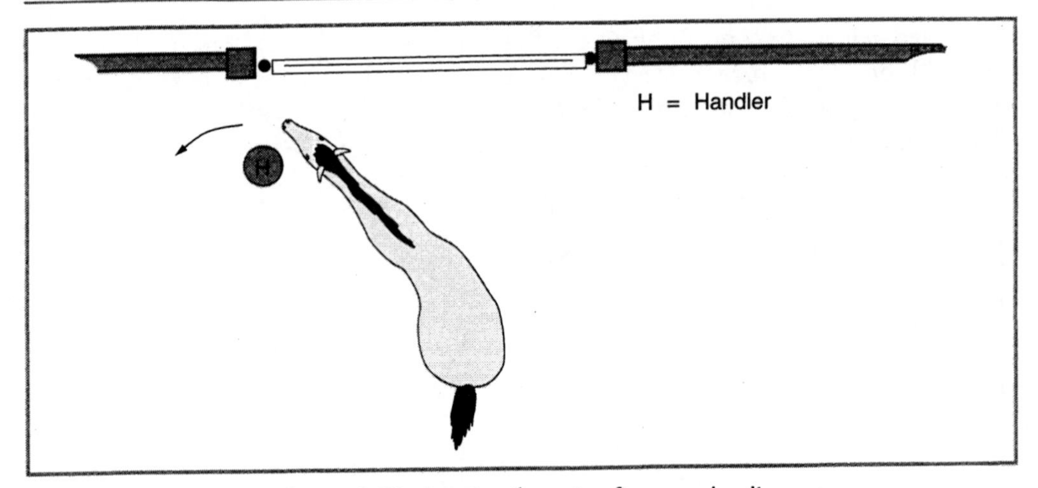

Figure 4-15 Leaving the gate after securing it

To return a horse to a paddock or pasture, you must again adapt your procedures to your situation. If there are other horses already in the paddock, you may need helpers or a whip to keep the other horses away from you until you complete the release. Open the gate and lead the horse through. Controlling the gate as well as possible, turn the horse around to face it. Remove the lead rope and quickly step back through the gate opening and close it again. This prevents the horse from accidentally hurting you if it spins around and kicks up its heels as it runs off (see Figures 4-16, 4-17, and 4-18). At some facilities such as breeding farms, it is not unusual to be

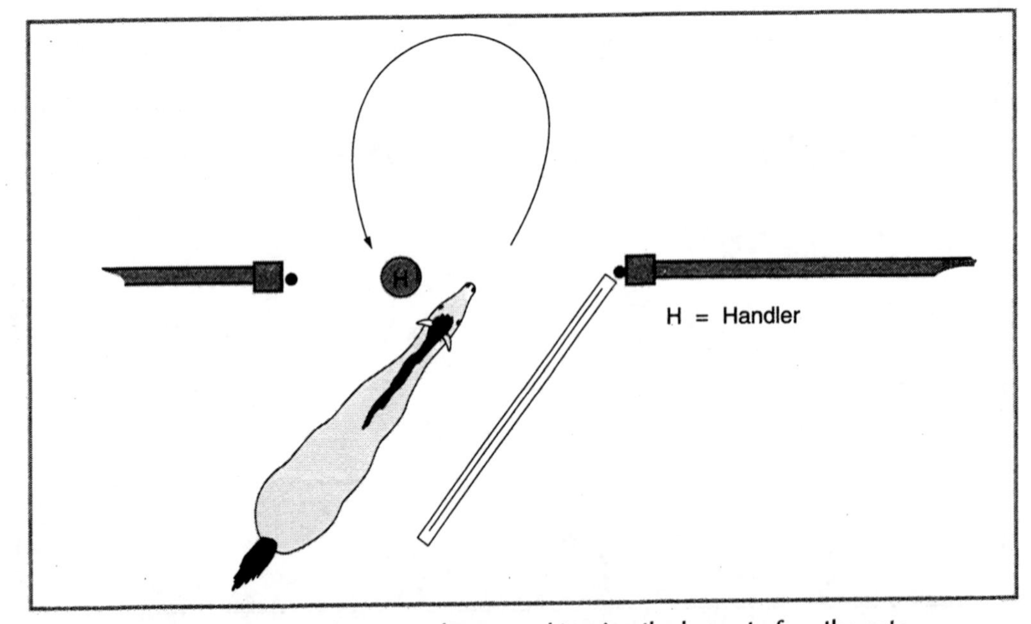

Figure 4-16 Entering an enclosure and turning the horse to face the gate

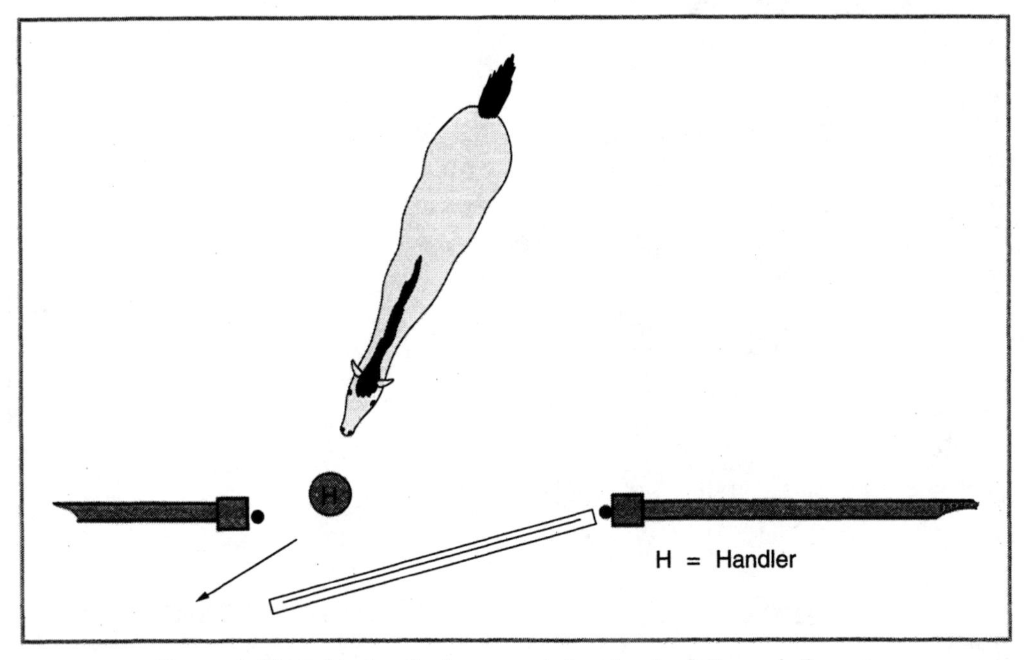

Figure 4-17 Releasing the horse and stepping back through the gate

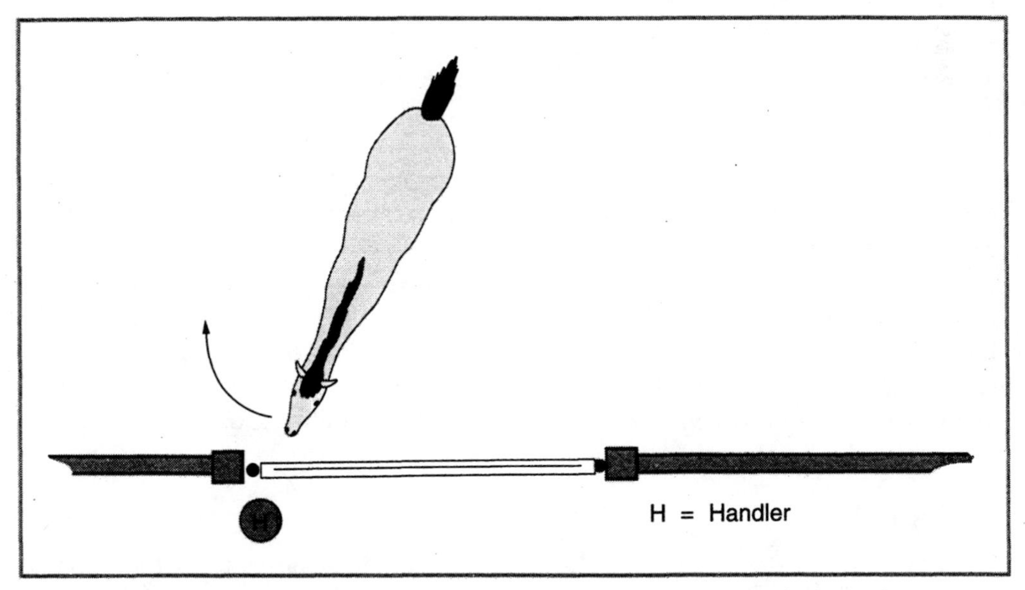

Figure 4-18 The handler safe on the outside of the enclosure as the horse leaves

turning out multiple horses simultaneously. In this case follow the same procedures, except be sure to line the horses up in a straight line facing the gate before releasing them. Then release them all at the same time so that each handler can step directly to the gate without being cut off by someone else's horse (see Figure 4-19).

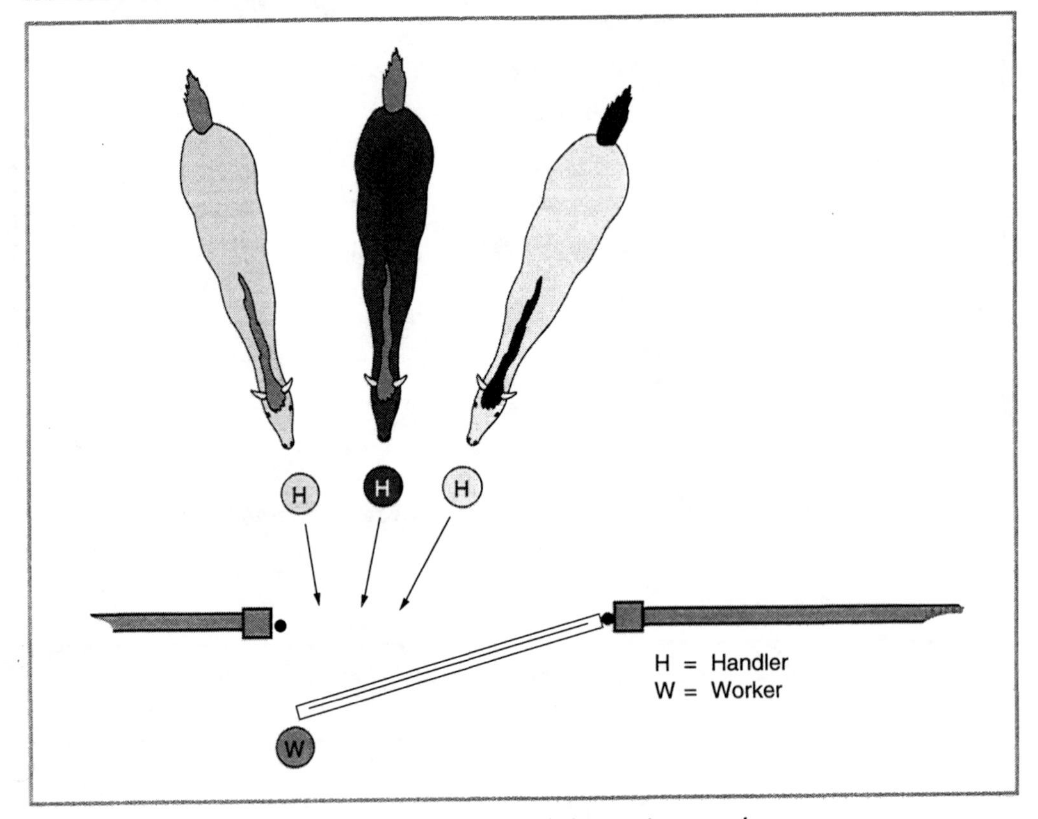

Figure 4-19 Releasing multiple horses in an enclosure

Working on a horse in a paddock or pasture is not difficult if the horse is alone. It can simply be held by another handler or tied to a secure fence post. Some people even rig crossties in the corners. If there are other horses in the paddock, there can be more of the playing and heckling mentioned earlier. For this reason, many people prefer to remove the horse from the paddock and work on it elsewhere, or remove all the other horses, leaving only the horse to be worked on. If these options are not possible, at least have some helpers keep the other horses away from you while you work. It is difficult to do good work in a safe manner with multiple horses milling around you.

SUMMARY

Working safely in stalls and paddocks involves making adjustments for the walls and fences. Handlers must not position themselves in such a manner that the horse can catch them up against a solid object, or cut them off from the nearest exit. It is tempting to lower one's guard with familiar horses and to accept poor positioning. Unfortunately this eventually leads to lowering one's guard with most horses, which is often when trouble begins.

Lungeing and Ground Driving

LUNGEING

For the purposes of this book, lungeing will be defined as an activity where the horse moves in a circle around the handler. Advocates of the Jeffery Method (an Australian method of training) have a slightly different definition for the word, but the majority of trainers think of circular motion around them when they use it. This is quite a consensus when you realize that we can't even get agreement on how to spell the word (there are at least four different spellings). Fortunately there is some agreement on the use of the activity. It is commonly used to give the horse physical exercise, to change the way it thinks, or both. Although some trainers feel that moving young horses in a small circle is not good for their legs, the majority of trainers see lungeing as a useful tool in the development of their young horses. We will divide lungeing into two categories: conventional and free.

Conventional Lungeing

Conventional lungeing is characterized by having a long line or rope (often called a "lunge line") connected to the horse's head, either by way of the halter or the rings of a special piece of equipment for the head called a "lungeing caveson." Trainers can hold on to the line and have some semblance of control as the circle begins to expand out away from them.

Before beginning a session, it is important to check the area in which you intend to work. There should be no dangerous ruts in the ground where you or the horse could trip and be injured. It is best if there are no sharp protrusions in the area, or other objects you or the horse might run into that would cause injury. If there are to be observers, there should either be enough room for them to move out of harm's way, or a safe area set aside for them.

The horse you are working with should be a safe match for you. If it is too unruly or inexperienced, you may find it best to go back to leading exercises where you have a shorter lead. You can then master its behavior using "in hand" techniques before moving on to lungeing. The shorter lead gives you more control and allows the horse less room to maneuver, giving you an advantage until such time as the horse gains the experience necessary to be reliable on a longer line. Your experience level will be an important factor in this decision. If you have many years of experience lungeing difficult horses, then you can probably handle animals that are too difficult for beginners to work with safely. If you are less experienced, don't work with horses that are beyond your present skill level.

When possible, carry a whip when lungeing. As the horse begins to move around you, it has the option of staying out on the circle or turning in towards the center. You are usually safer with the horse out on the circle, because a horse can't hurt you (accidentally or otherwise) if it can't reach you. You can't push the horse away with the line, you can only wave the excess line at it (if there is any at the time you need it). A whip is better at encouraging the horse to stay away from you. Without it, you are less safe. Unfortunately, whips are controversial. Many people do not use them because they are reluctant to inflict pain (or to appear as someone who enjoys hitting horses). I have no quarrel with this line of reasoning. There is nothing wrong with being reluctant to inflict pain—I certainly am. However, there is a difference between being reluctant to hurt other creatures and not carrying a whip for emergencies. Even the best horse can become frightened and run you over trying to escape if you are not carrying something to fend it off. You do not have to become a whip-wielding warmonger, you are simply taking a reasonable precaution. How you use the whip, if at all, is your business. A good rule of thumb is to use it only when necessary and not to enjoy it even then.

Each year I am told by at least one student that he or she can't use a whip with a particular horse because the horse is afraid of the whip. Rarely is this true. In the few cases where it is, our time is much better spent desensitizing the horse (teaching it not to fear whips) than it is training with an animal whose fear dictates what tools humans are allowed to use. What tools we use should be our decision; it is not safe to have the horse dictating our choices.

It is not unusual to start the horse with small circles where the line is short and the horse close to the trainer. At these distances it is possible for the horse to swing its hind end in toward the trainer, putting the human in Zone D at close range (see Figure 5-1). Controlling the horse's head is the key to avoiding this situation. Remember that the rear end of the horse moves in the opposite direction from the head, so keeping the head straight or turned in should keep the rear end straight or pointed out away from you (see Figure 5-2).

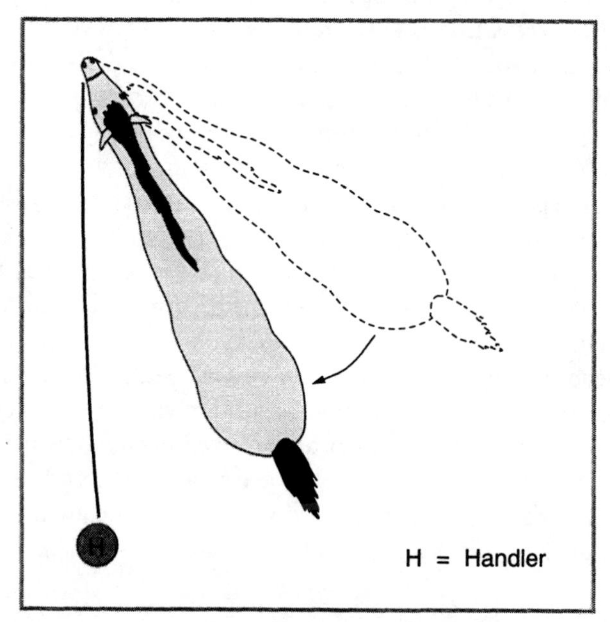

Figure 5-1 Not controlling the head puts the handler at risk.

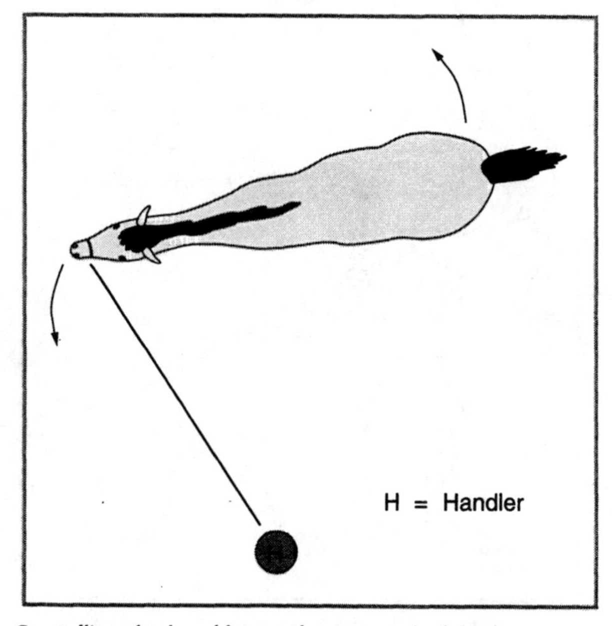

Figure 5-2 Contolling the head keeps the rear end of the horse at a safe distance.

At these short distances you do not need all the line, creating excess on your end. You should not let the excess dangle on the ground where it can wrap around your feet and legs as you turn (you will find it most difficult to lunge a horse properly when you are lying on the ground with your legs all tied up). Opinions vary as to which method of dealing with the excess line is best, but most versions have it held in the trainer's hands, either folded or coiled in loops (see Figures 5-3, 5-4, 5-5, and 5-6). Whichever method you employ, do not wrap the excess around your hand or tie the line to any part of your body or clothing to keep the horse from getting away. There are times when, for your own safety, you need to release your hold on the horse completely—for instance, if the horse panics, manages to turn its body to face away from you, and bolts forward. If it is truly panicked, it may be in the next county before it stops to think. There is very little you can do after an animal that size is facing away from you, because it is extremely powerful moving in the forward direction. Your choices are either to let go and catch it again later or to be dragged into the next county on the end of the line, flopping like a dead fish on the end of a long rope. Needless to say, this is a good time to let go. Wrapping the end of the line around your hand either slows down or completely inhibits your release and is often referred to as the "suicide grip." Tieing it to your body is even worse. But if you hold the line properly, and use the whip to keep the horse out on the circle so that the line doesn't touch the ground between you and the horse, things usually go well. They at least go safely (see Figure 5-7).

Figure 5-3 Excess line held in folds in the lead hand

Figure 5-4 Excess line held in folds in the whip hand

Figure 5-5 Excess line held in coils in the lead hand

Figure 5-6 Excess line held in coils in the whip hand

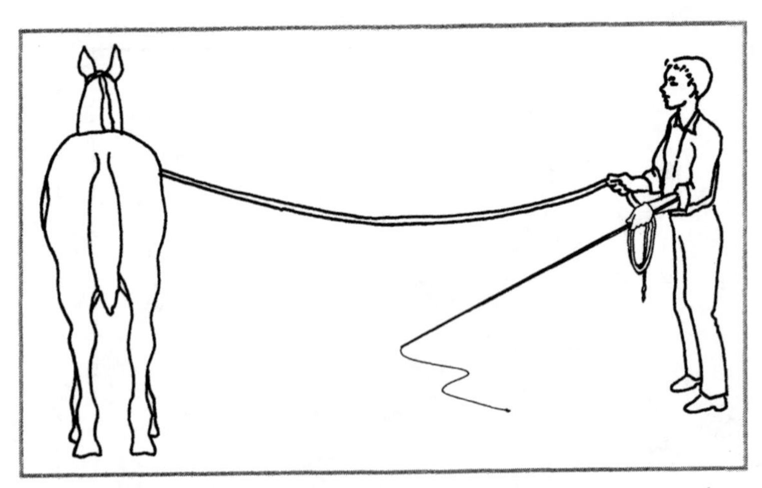

Figure 5-7 Horse at a safe distance and line controlled properly

Horses react to startling stimuli differently. Some do not maneuver to run away, as described earlier, but rather stay in the circle, increasing their speed until they are moving so fast that they are completely out of control. Should you be run away with like this, try to remember the relationship between speed and the size of the circle. When the circle is small, it is difficult for the horse to go fast because the curve is so sharp. Most horses choose to go slowly when the circle is small, which is why many trainers will tell you to give the horse more line (increasing the size of the circle) when you ask it to go faster. When the horse is going so fast that it is out of control, a good approach is to reel in the line, deliberately making the circle smaller. It may take some strength to do this, but horse training is a physical game. As the circle gets smaller, you will reach a point where it is so small that the horse will decide to slow down. When it does, you can reestablish control.

It is good to face the horse at all times while lungeing. Some people stand still and allow the horse to move around behind them. As it does, they pass the lunge line over their head and wait for the horse to complete the circle and show up in front of them again. If something happens to frighten the horse while it is behind the handler, he or she is hardly in a position to handle the situation. Even the best horse can become frightened, so try to develop the habit of facing the horse.

If the horse you are working is a dominant type of animal who thinks highly of itself, don't push it far enough to start a fight. There are times when dominant horses resist on certain tasks. Sometimes they are merely testing to see if your rules are still the same, or if you are willing to give up the leadership role. In these instances firmness and patience on your part will bring them through. On the other hand, there are times when these animals simply refuse to do certain tasks in a manner that is much more determined. When they are in this frame of mind, they do whatever is necessary to compete with you, and lungeing is not the safest way to handle them. You have much less control of the animal during lungeing. The horse has the room and freedom to maneuver, which gives it the advantage during any confrontation. If your dominant horse is being difficult, it is safest to avoid conflict by choosing to do things the horse is willing to perform. It then has no reason to act poorly. When you wish to work on the items it is resisting, it is safer to switch to leading the horse on a short lead, using what are known as "in hand" techniques where you have more control. When you have changed the way the horse is thinking, you can safely return to the lunge line.

Free Lungeing

For the purposes of this book, free lungeing is described as a process where the horse moves around the trainer while there is no line or other equipment forming a physical connection between the two (see Figure 5-8). The name comes from the fact that the horse is free while lungeing. This process has also been called "tackless training," "lungeing at liberty," or "liberty training." Many trainers feel that it is the most powerful method of changing the way a horse thinks, whereas others use it merely for physical exercise.

As with conventional lungeing, good free lungeing starts with a safe work area. The footing should not have any deep ruts or obstructions to trip the human or the horse. If you are fortunate enough to have a round pen in which to work, the wall should be sturdy. Some horses challenge the wall to get out of the trainer's space, and the wall needs to be sturdy enough to withstand this. The wall should be high enough to discourage jumping, and there should be no spaces in it large enough for a horse's

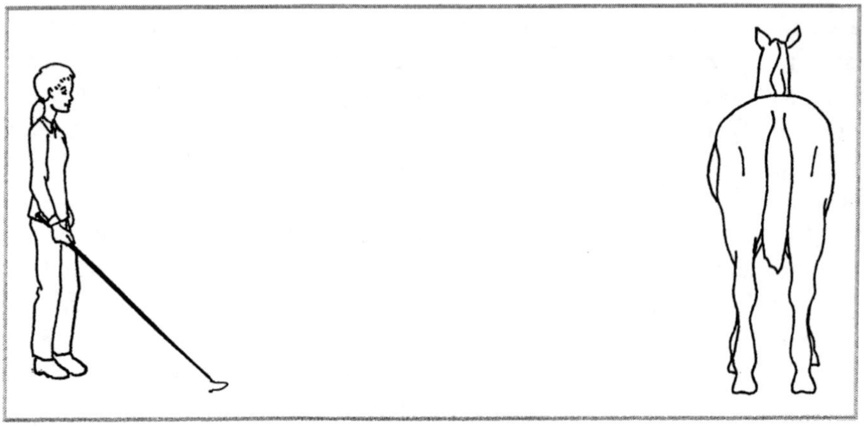

Figure 5-8 Free lungeing

hoof to pass through. Adequate lighting is a benefit, because much of free lungeing involves visual communication between you and the horse.

The suitability of the horse for the human is again important. Do not work animals that are beyond your present skill level.

To carry a whip is as important for free lungeing as it is for conventional lungeing. It is still the best way to encourage the horse to move away from you, and a horse can't hurt what it can't reach. If the horse shows a willingness to kick up its feet, get a longer whip and stand farther back than usual. For the reasons just mentioned, your first priority when you begin free lungeing is to defend your personal space. The horse should not come into your space unless you are inviting it to. This is also important because this type of lungeing is a form of herding, and you cannot move a horse around the pen if you cannot keep it out of your space.

Do not put too much pressure on the horse while you are working it. Read the horse carefully for signs of stress, and always be sure that the horse has somewhere it can move to release pressure. If you exert too much pressure, some horses feel forced to break the wall to escape. Others feel forced to resist or fight you, which can create an unsafe situation. As with conventional lungeing, use subtle methods to deal with dominant or difficult horses. It is not safe to go around picking fights. Fortunately these subtle methods are extremely effective and quite safe.

If more than one person is working the same horse, the procedure you use to switch trainers is important. One method that has worked well is to have the first trainer invite the horse into the center of the ring or catch the horse in some manner (for instructions on how to invite the horse in, see my book *Fundamentals of Free Lungeing*). While the first trainer holds the horse in the center of the ring, the second trainer approaches safely and takes control of the horse. While the second trainer holds the horse, the first trainer exits the training area. Only after the first trainer has safely exited does the second trainer release the horse. To do so sooner creates an unsafe situation where the first trainer is trying to get out and the horse is loose to create mischief (see Figures 5-9, 5-10, and 5-11).

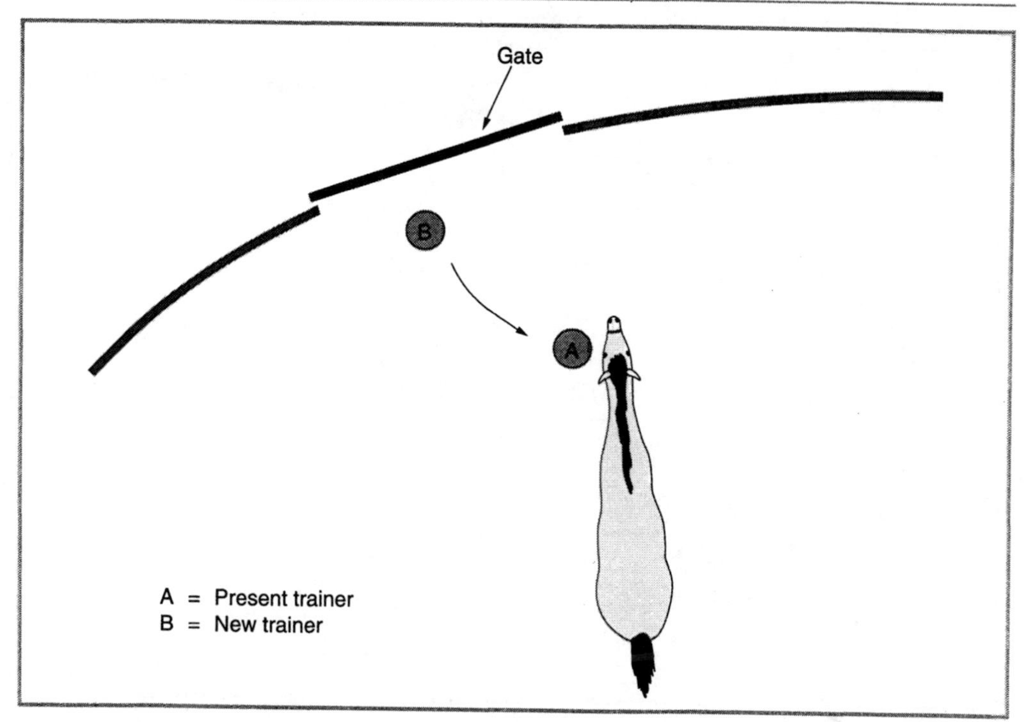

Figure 5-9 The present trainer *(A)* holds the horse while the new trainer *(B)* approaches.

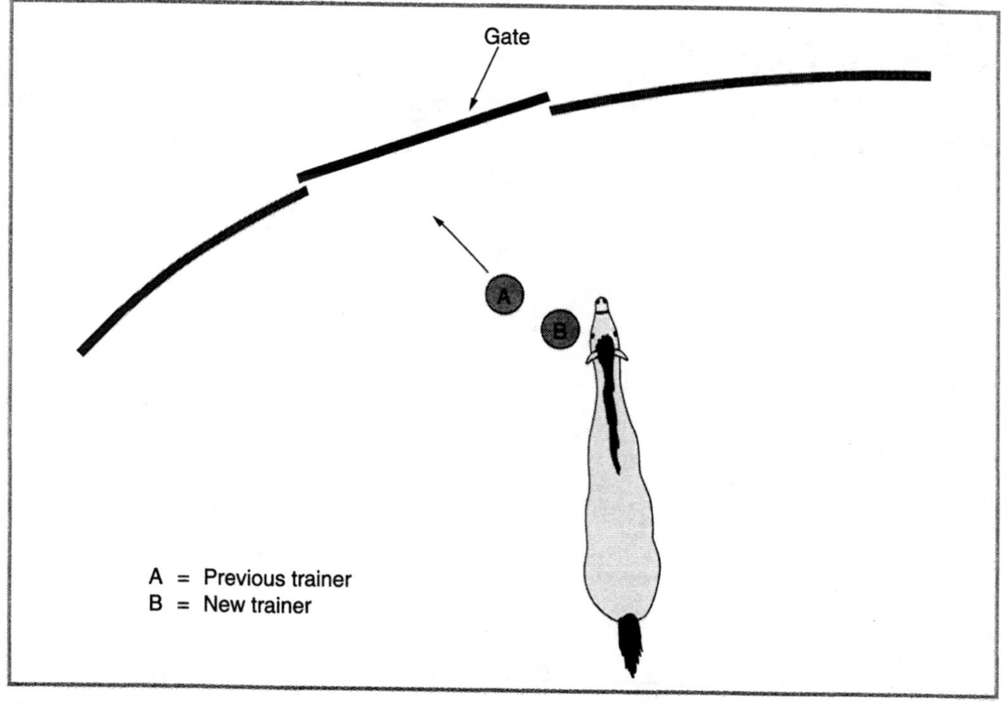

Figure 5-10 The new trainer *(B)* holds the horse while the previous trainer *(A)* exits the training area.

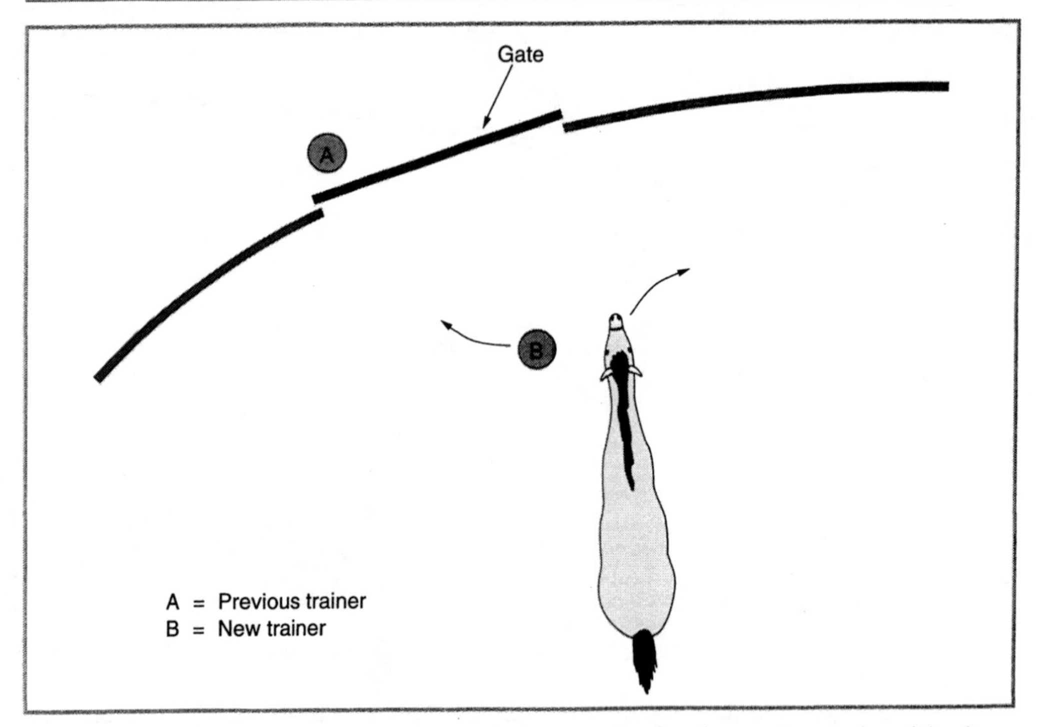

Figure 5-11 The new trainer *(B)* releases the horse only after the previous trainer *(A)* exits.

Ground Driving

Ground driving, often referred to as long lineing, is a process where the horse can move in a starighht line as well as a circle. Usually the horse is in front of the trainer and equipped with a bridle complete with a bit in the mouth and long lines connecting the bit rings to the trainer's hands (see Figure 5-12). The first name comes from the fact that the trainer walks along behind the horse and drives much like the driving of horse

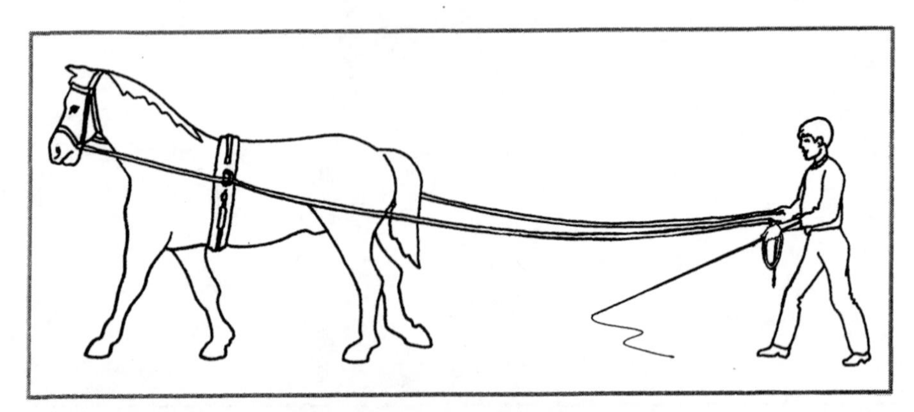

Figure 5-12 Long lineing

drawn vehicles; the second from the length of the lines connecting the bit with the trainer's hands. Not all trainers use this technique, but the ones who do feel that it helps teach the young horse what the bit cues mean and develops a soft mouth so that the horse responds to very subtle stimuli from the bit.

The first thing to remember when ground driving is that because you must work in Zone D, your choices must be wise ones. Don't attempt ground driving until the horse has proven its ability to work calmly and safely in all other procedures. If you are not certain of the horse's responses, choose methods that have another person holding its head in the early stages. There are probably several ways to do this, but one that works well is the use of a third line. This is a method where a second person controls the head until the trainer and horse are moving well, and then backs away from the head holding a third line, usually connected to the halter. This way, if the horse becomes frightened or starts to misbehave, the trainer can drop the two lines connected to the bit, which saves the horse's mouth from possible injury. The two humans can then concentrate on keeping a hold of the third line until the horse calms down or starts behaving again (see Figure 5-13).

Regardless of how you choose to get things started, the distance from you to the horse is important. There are only two ways to be safe when you are working directly behind the horse: right up next to the buttocks or back so far that the extended legs can't reach you. In the first case you will be so close to the horse that should it kick,

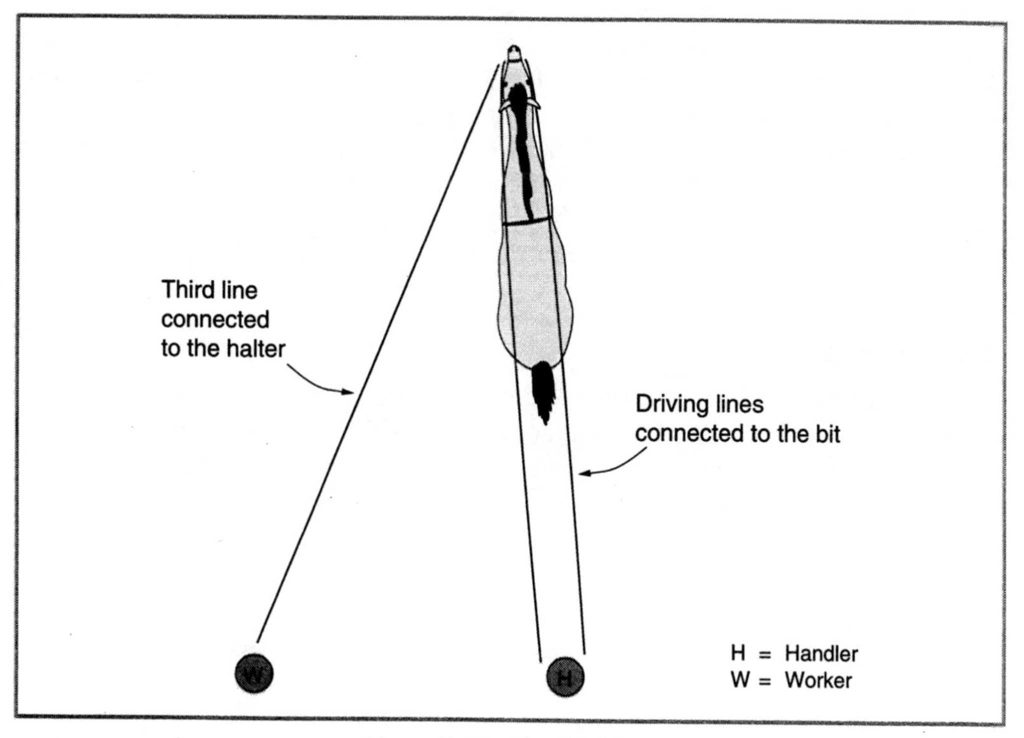

Figure 5-13 The third line

the hooves will not have space to gain their full momentum before they make contact with you. This jams the kick and robs it of its power. If you get kicked, you will experience pain, but the chances of being seriously damaged are small. They are even smaller, of course, if you are smart enough to stay out of reach completely. The complication to this is that the farther back you stay, the more difficult it is to control the horse. So you must make a compromise, but make it a safe one. Always remember the famous question—how long are a horse's legs? Answer—longer than you think. As the old adage says, it's better to be safe than sorry.

Dealing with the excess line is more complicated now because the trainer has to handle two of them. Many trainers deal with them the same way they deal with lunge lines, by folding or coiling the excess. Others simply let the excess amount trail behind them on the ground, which works due to the fact that the trainer is moving forward, not standing in the middle of a lungeing circle. If you choose to coil or fold the lines, make sure that you practice making the lines longer and shorter. It's not as easy as it sounds. Good line skills and using your feet properly keep the lines from getting too much slack in them. If they become too loose, they can sag and get stepped over or even wrapped around the horse's legs. Either of these can frighten the horse and create unsafe conditions.

A whip is just as important while ground driving as it is any other time. I was once told that horses couldn't move well to the rear. Unfortunately, someone forgot to tell the horses. They are certainly not as efficient going backward as they are going forward, but they are quick enough to cause you trouble, especially if you are unprepared. If you always carry a whip, the horse develops habits, which do not include going backward. Unfortunately, trying to manipulate two lines and a whip with only two hands can be an infernal nuisance until you have practiced it enough. This is probably the major reason many people don't use the whip. It's one more thing to lug around and extremely difficult for beginners when combined with the two lines. However, when you need the whip, it's often too late to go back and get it. The one time you need it more than makes up for the nuisance factor.

SUMMARY

Lungeing safety is largely a matter of having a safe work area, having the correct equipment, working horses of the appropriate level of difficulty, and being able to keep them at the correct distance. The same is largely true of ground driving. It is difficult for horses to injure things they cannot reach, and few tools keep them at a distance as well as a good whip.

Tack

Condition of Tack

The assorted equipment that connects us to the horse and helps us to manage and control it is often referred to as "tack." For example, items such as halters, bridles, saddles, leads, and lunge lines fall into this category. Because their purpose is to help us control the horse, it is better for us when they work correctly. Failure or breaking of tack can lead to unsafe situations. Therefore, the condition of your tack is something you need to monitor carefully.

Tack is made of different materials, from leather and nylon to more-recent synthetics. Each has its own strengths and weaknesses, and you should be aware of these for each piece of tack you use. For instance, leather halters are difficult to clean and maintain, but they break when a horse panics and throws itself back on the crossties. This allows the horse to break free and calm down, which is often a safety advantage both for the horse and the human. On the other hand, nylon halters are simple to clean and maintain but are much stronger than leather halters and do not usually break when the horse panics and throws itself back on the crossties. Consequently the horse may not break free and may become even more frightened and violent. In this scenario it is safer for the material to break, but usually it is best for tack to be strong and not give way. Each material has its pros and cons, and all can be made to work safely. It is your responsibility to know the characteristics of the material you are working with and adjust procedures to provide the safest possible conditions. In the previous example, people who use nylon halters must be particularly careful about the safety releases on their crossties and other tieing points because they are not going to get any help from their halters if a horse panics.

Materials can also be in poor condition. Stitching can rot and become weak. Because this is what holds most of the piece together, the condition of the stitching should be carefully checked before using any piece of tack. Leather can also rot, dry, and split, whereas nylon can fray and give out. This is particularly true in areas where the material bends sharply to pass around metal rings, and so on. So it is good procedure to inspect tack to see that it is in good condition and that the stitching is sound before use.

Lead and Lunge Lines

Lead lines should be inspected to see if there are any cracks or evidence of weakness in the snap or clip used to connect it to the horse's halter. Remember that snaps are only as strong as the small pin that connects the two portions of metal (see Figure 6-1). The metal pieces used in today's tack do not seem as strong as they did in years past, so it is best not to expect them to stand up to too much strain. In an emergency, lead material is quite strong and can do many useful things, but remember that the snaps are designed to lead horses, not pull trucks out of ditches. While you are looking at the snap, check the lead material where it wraps around the base of the snap. It should not be dry, split, rotted, or fraying, and the stitches should be in good condition. The length of the lead should be appropriate to the horse being handled and the situation. Difficult horses should be handled with longer leads (often called "stallion leads") so that when they rear or act in other foolish ways, the handler can let out line without having to release the horse completely. Keep in mind that a lead is only as strong as its weakest point, so check the entire length of these longer leads. Whether or not to have a knot in the end to hold on to is a controversial subject. Some people keep a knot in the end

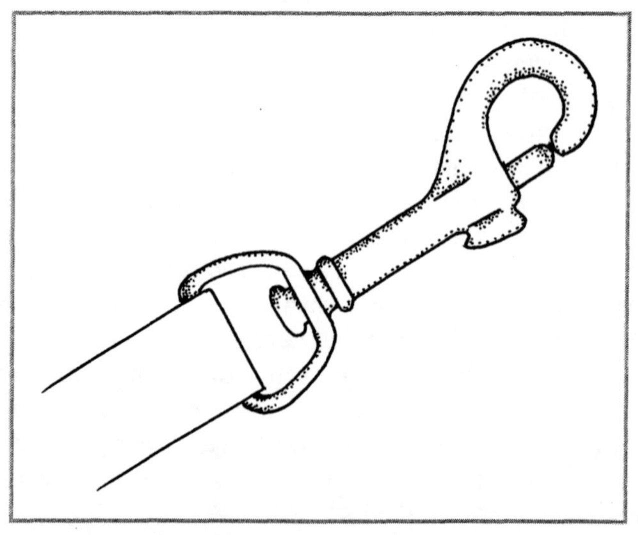

Figure 6-1 Snaps are only as strong as the pin.

of their longer lines and have had no problems. Others claim that when the lead gets pulled out of their hands in difficult situations, the knot creates injuries as it flies through the hands at high speed. It is probably safer to have smooth ends to your longer leads rather than to knot them. Lunge lines are essentially even longer leads and should be inspected in the manner just described. Many include a hand loop on the non-snap end. The stitching and the material at the end of the loop should be inspected before using.

Halters and Bridles

The metal parts of a halter should be checked for cracks or weaknesses, and so should the material that bends sharply around the rings and buckles, if any. The inside surfaces that make contact with the horse should be smooth and without foreign objects such as burdocks, and so on. Anything that causes pain encourages the horse to panic and try to escape from it.

The metal parts of a bridle should also be free of cracks and weaknesses. They too need to be checked where the material bends sharply to pass around the metal pieces. Two locations that deserve particular attention are where the cheek piece connects to the bit and where the reins connect to the bit (see Figure 6-2). The width of the reins needs to be appropriate for the younger riders. Some small hands have difficulty controlling ¾-inch reins and do better with ½-inch reins. Both riding and driving reins or lines should be checked periodically by tying them to something solid and pulling on them to see if they are still strong enough to serve well.

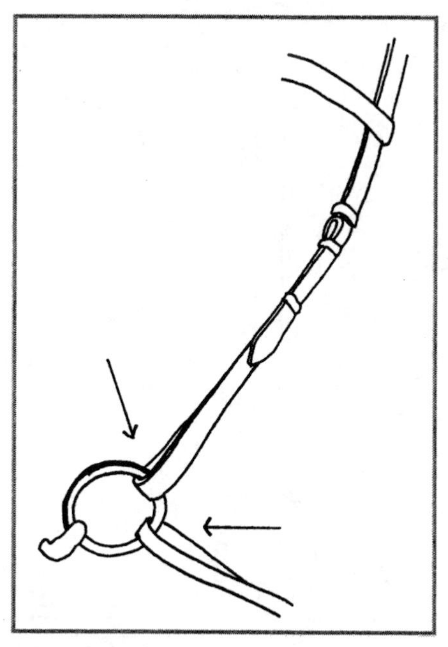

Figure 6-2 Cheek pieces and reins should be checked regularly.

Equitation Saddles

English Saddles. An area of concern with English saddles is the stirrup leathers. These are the strips of leather that hold the metal stirrup and connect it to the saddle. They are best made with triple stitching and a layer of nylon between the layers of leather to provide extra strength and reduce stretching. Naturally these stitches and materials should be in good repair, especially where they bend around the buckles and stirrup irons (see Figure 6-3). The stirrup bar is designed to allow the stirrup leather to separate from the saddle when the rider hangs a foot up in the stirrup iron during a fall. Theoretically this prevents the rider from being dragged after the fall (dragging can create extremely serious injuries). This sounds good on paper and in fact has worked well in many instances. However, there have been cases where the bar has failed to release under pressure. It is therefore safer to leave the bar down rather than up (see Figure 6-4). Another device to prevent being dragged is the peacock stirrup, which has elastic for the outside bar of the stirrup iron. This allows the foot to rip out of the stirrup under pressure. Boots with good heels help reduce the chance of the foot passing through the stirrup, However, most boots worn by English riders have only a half-inch heel, which is minimal at best. Basket stirrups prevent the foot from passing through the stirrup regardless of the style of boot worn or the size of the heel (see Figure 6-5). This helps reduce the chance of being dragged. But the stirrup is being accepted slowly due to its nontraditional appearance.

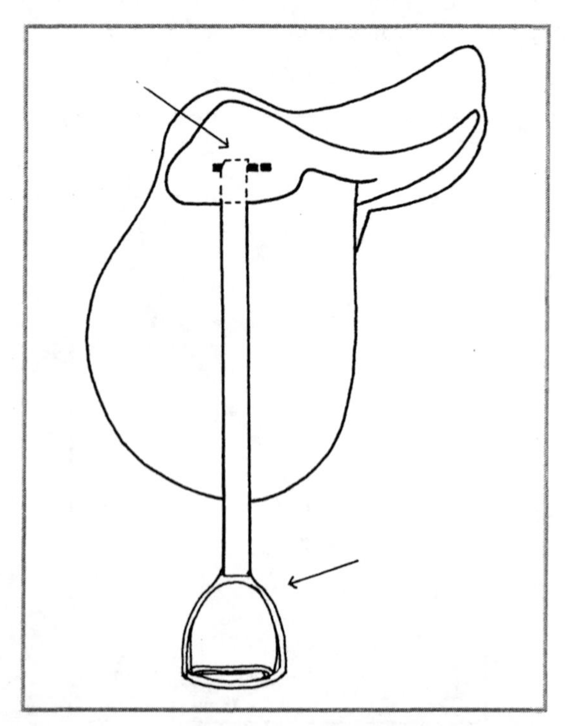

Figure 6-3 Inspection sites for stirrup leathers

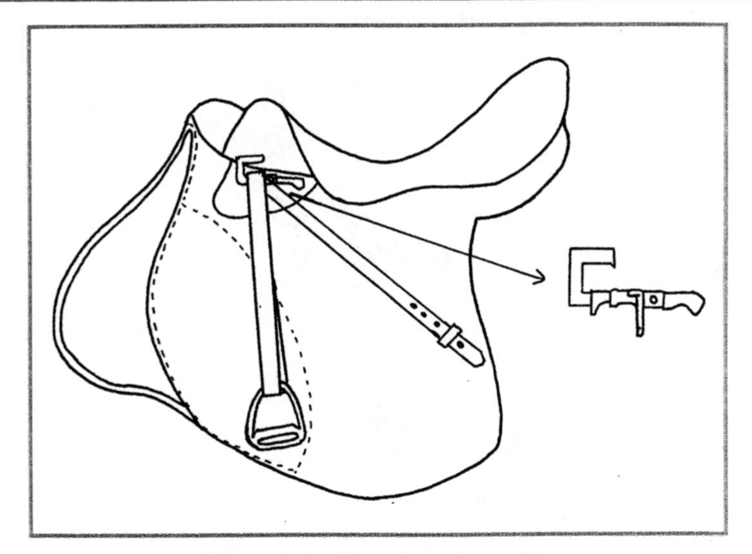

Figure 6-4 The stirrup bar safety

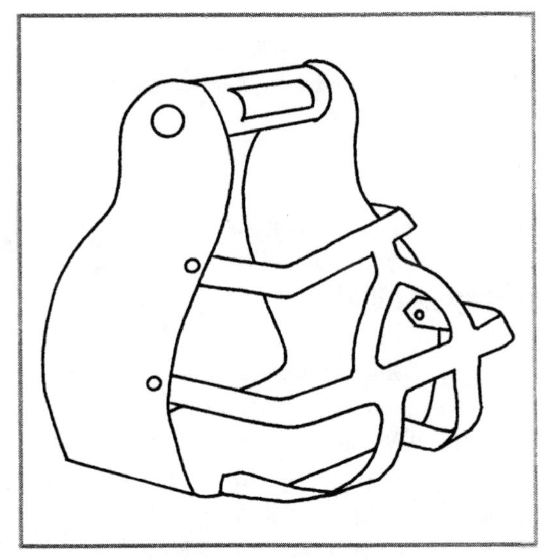

Figure 6-5 The basket stirrup

The billets (the strips of leather that connect to the girth buckles) often connect to nylon straps affixed to the saddletree. Usually there are three billets, the first being connected to the forward nylon strap and the second and third both connected to the rear strap. This is why so many instructors have students connect the forward buckle of the girth to the first strap and the rear buckle to either the second or third billet. That way, if one nylon strap gives way, the girth will still function.

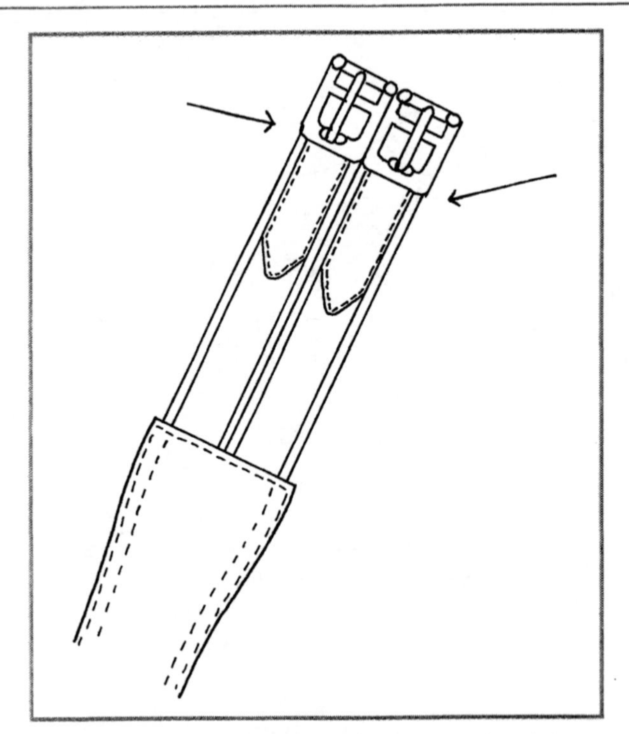

Figure 6-6 Potential trouble spots on the girth

The girth can also have any of the material or hardware problems mentioned earlier, so it pays to inspect it for metal weaknesses or material problems, particularly where the leather wraps around the buckles (see Figure 6-6).

Western Saddles. Because the stirrup leathers are not removable on most western saddles, their condition and fit are of particular concern. Stirrups should be at least ¼ inch wider than the width of the boot being worn by the rider. This helps reduce the chance of the boot hanging up in the stirrup. Peacock stirrups are not common in western saddles, but basket stirrups work well and are recommended. If these are unavailable, boots with long heels are recommended to prevent the foot from passing through the stirrup. The position of the stirrup leathers can vary in western saddles, some forcing the rider to extend the leg forward of the hips, making it difficult for beginners to maintain balance. It is often safer to start beginners with saddles that have the stirrup leathers hung like an English saddle, such that the leg can be comfortably positioned with a straight line being drawn from the rider's head through the hip to the heel (see Figure 6-7).

The cinch should be inspected as carefully as the girth on an English saddle. The condition of the rings and latigo leathers that connect the cinch to the saddle need to be in good repair and tied properly.

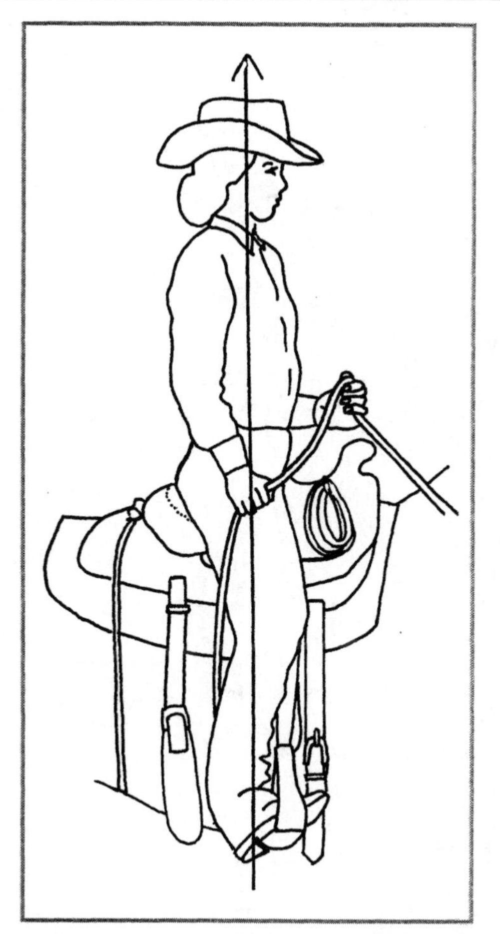

Figure 6-7 Western saddle set up to maintain beginner's natural balance

Driving Saddles and Harness

Driving saddles, or pads as they are called in some harnesses, need to have their stitching and hardware inspected as described earlier. The extra pressure exerted on the harness when the cart or carriage must be drawn up or down steep inclines makes the inspection of the hardware, leather, and stitching even more important than usual. The tongues of buckles have been known to bend to the point of pulling through their frames, releasing the straps they were supposed to keep connected. For this reason, it is often recommended not to use brass buckles on straps that will be subjected to great amounts of pressure (for instance any buckles associated with the hames, breast collar, traces, breeching, or breeching straps). The metal hames, which rest in a groove in the neck collar and help transfer forward energy from the horse to the carriage, need to be made of steel rather than softer brass if heavy loads are drawn. Hardware on the carriage or cart itself needs to be inspected also (see Figures 6-8 and 6-9).

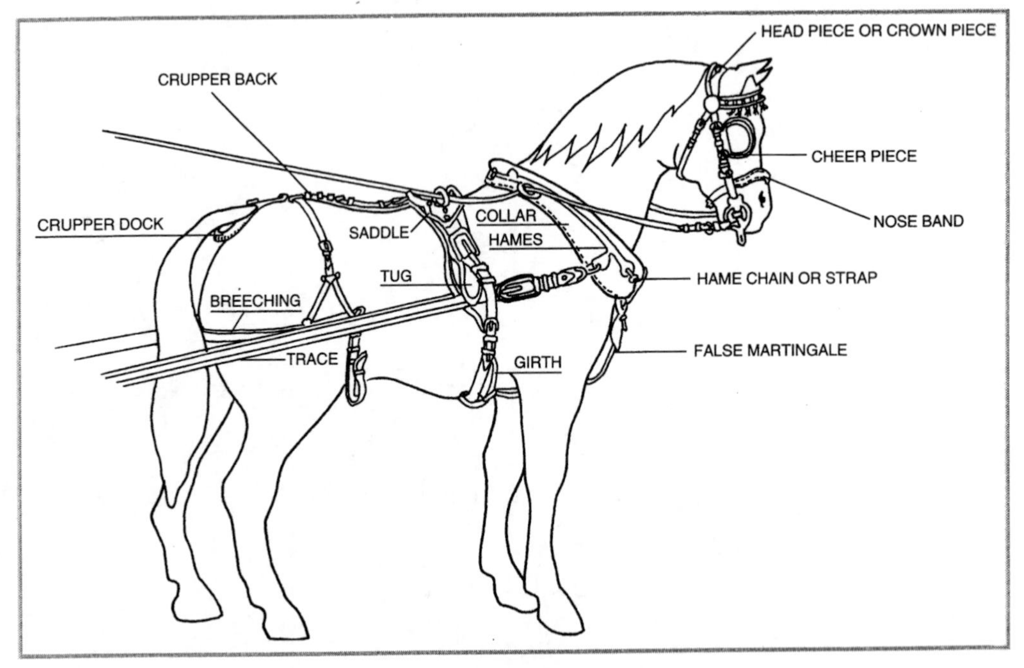

Figure 6-8 Harness parts

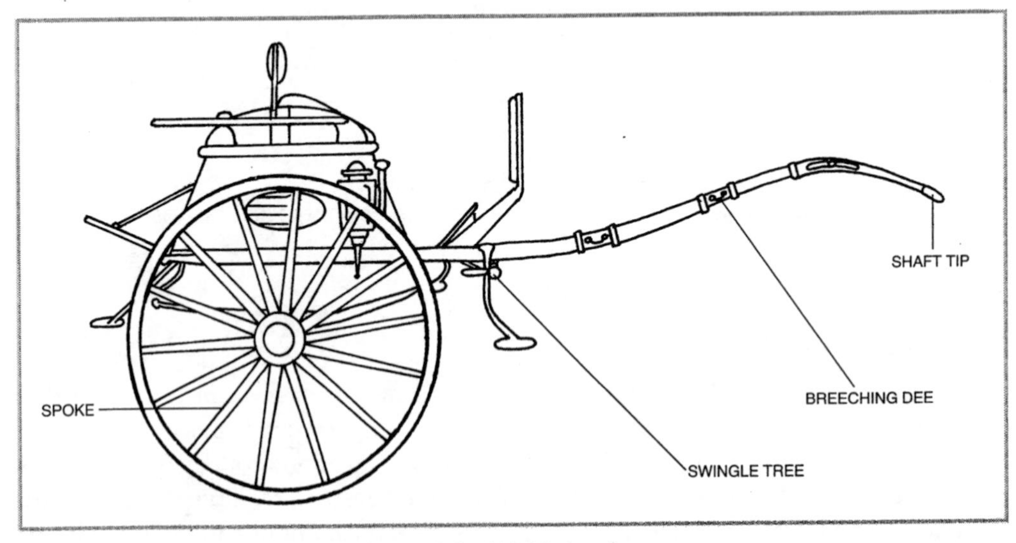

Figure 6-9 Vehicle hardware

How Tack Fits

Pain often creates bad behavior in horses, frequently to the point of being unsafe. Because poorly fitting tack can create pain, you are usually safer with tack that fits properly.

Halters and Bridles

The browband should not be too tight, because this will pinch the ears. Generally it is best to be able to put two fingers under any part of a halter or caveson to prevent pinching anywhere in the face and pressure on the sensitive area of the poll just behind the horse's ears (see Figure 6-10).

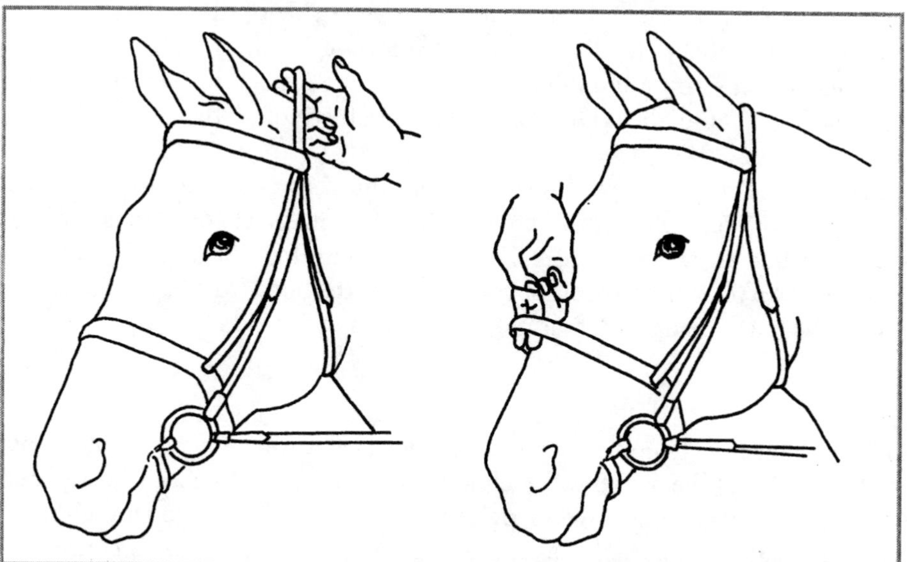

Figure 6-10 The two-finger test for bridles

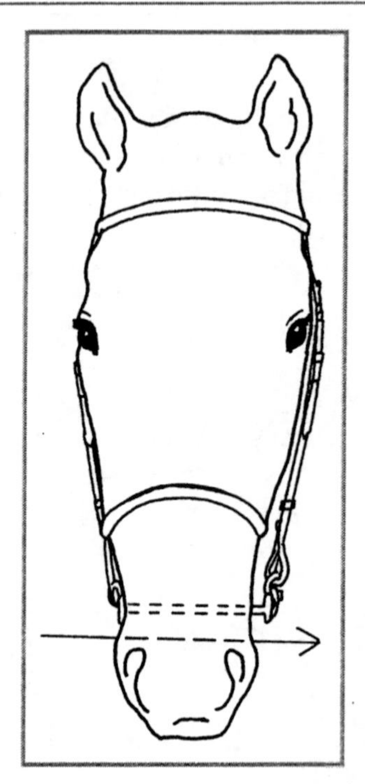

Figure 6-11 Bits should not project more than ⅛ inch when pulled to the side.

The bit is the piece of metal designed to be carried in the horse's mouth. When present (hackamores and bosals, for instance, do not have bits), this should be checked for sharp spots due to rusting or mechanical damage. It should also be the correct size for the horse. This can be checked by pulling on one side of the bit until it is snug on the opposite side of the mouth. In this condition, it should not stick out more than ⅛ inch on the side you are pulling towards (see Figure 6-11). Even the most gentle snaffle can bend at the hinge, striking the horse in the roof of the mouth when it is too large for the horse.

Driving bridles should be adjusted so that the horse's eyes are positioned in the middle of the winkers, blinkers, or blinders, whichever you choose to call them (which prevent the horse from seeing to the sides or the rear). The noseband should be adjusted to keep the winkers from separating from the head (to prevent the horse from seeing to the rear), being sure that nothing is rubbing the eyes.

Saddles

All types of saddles (riding or driving) should have a tree wide and high enough to prevent pinching of the withers where the horse's neck joins its back (see Figure 6-12). The tree should also be narrow enough to keep the gullet from touching the horse's spine (see Figure 6-13). As far as placement on the horse's back is concerned, the saddle should be in a position to allow one hand's width between the front edge of the

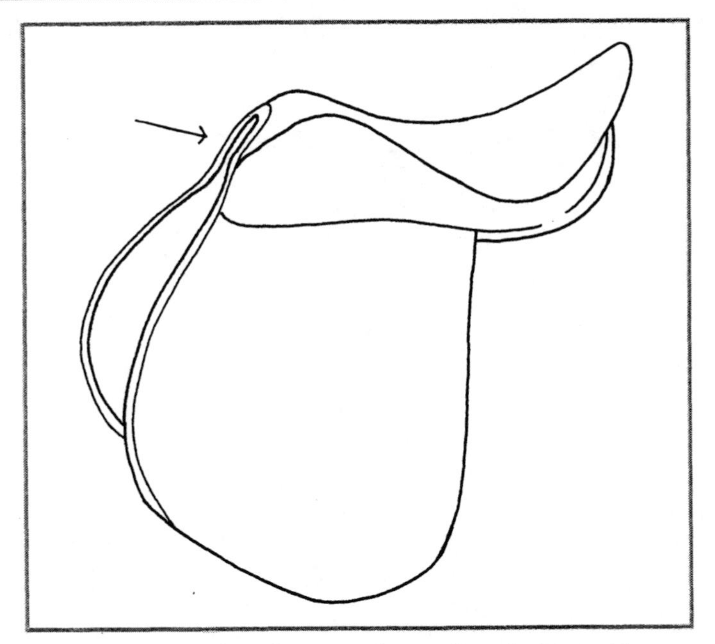

Figure 6-12 Area of concern for saddles pinching withers

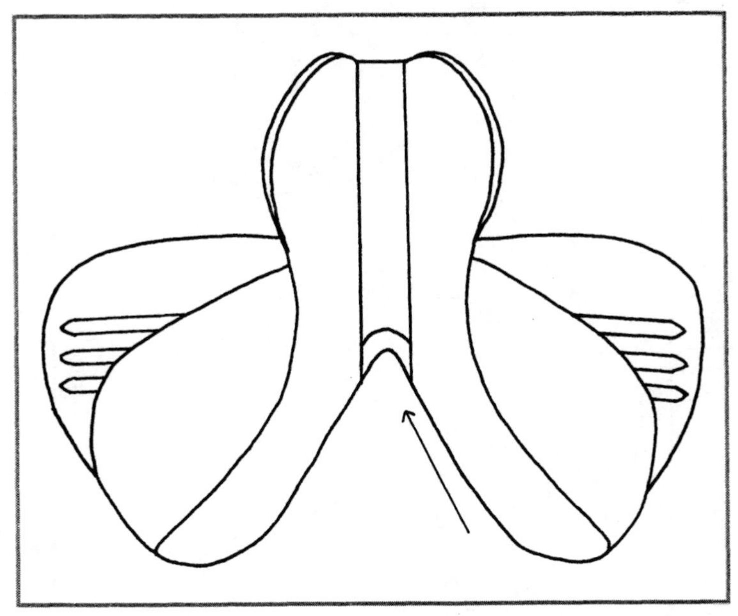

Figure 6-13 The gullet as seen from the underside of a saddle

girth and the horse's elbow (see Figure 6-14). The lowest point of an equitation saddle should be in the middle of the saddle (see Figure 6-15). When dealing with young riders, the length and width of a riding saddle should be appropriate for the physical

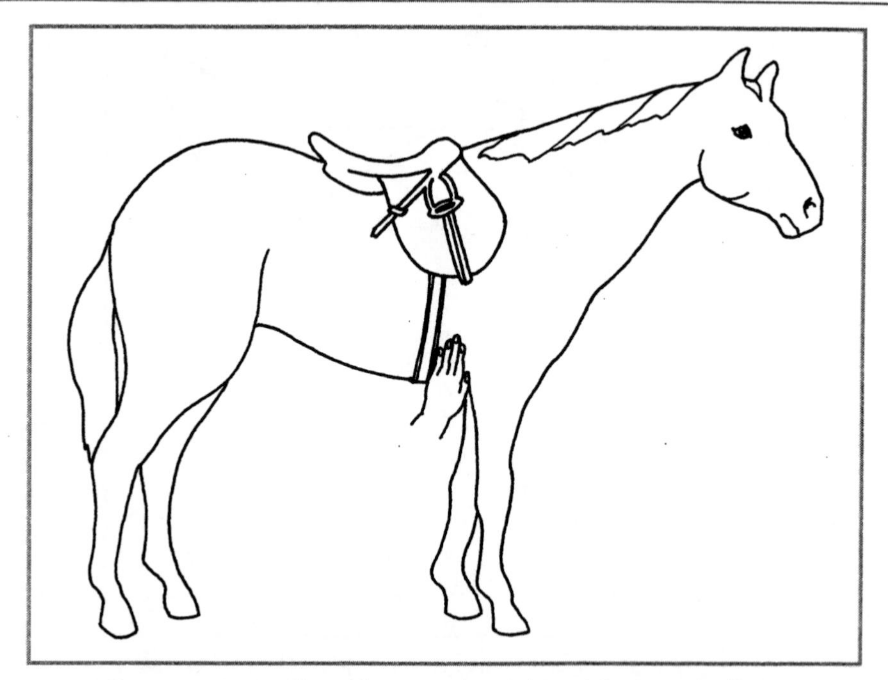

Figure 6-14 One hand between the girth and the horse's elbow

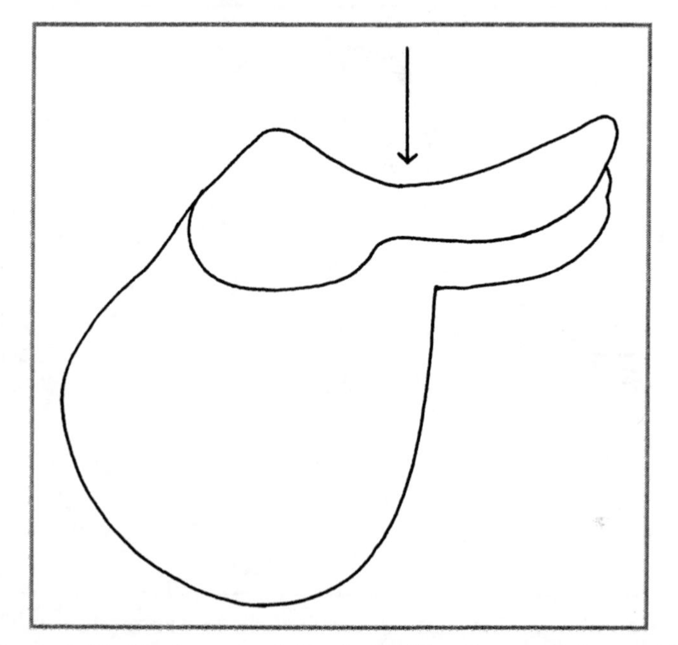

Figure 6-15 The lowest point in the middle of an equitation saddle

size of the person, otherwise it is difficult for these riders to put their legs on the horse and also to stay in the saddle if things go wrong.

Driving Harness

Before discussing the harness, it is good to recall that nothing is more important than the fit of the horse itself to the vehicle being drawn. To match a horse with a vehicle that is either too large or too small is foolish and should be avoided.

Starting at the rear end of the horse, the breeching strap helps the horse slow the vehicle and stop it. The strap should be sitting horizontally about halfway between the hocks and the point of the buttocks, and you should be able to put the width of your hand under it (see Figure 6-16). The crupper, the padded loop that fits under the horse's tail and is connected to the saddle by the backstrap, should be loose enough to place the width of a hand between the backstrap and the horse's loins (see Figure 6-17).

The tugs are leather loops that the shafts of the vehicle pass through. They should be adjusted to hold the shafts so that there is a slight rise as they go from the vehicle to the horse. The traces, or straps that connect the collar or breast collar to the vehicle, should be adjusted so that the tugs are resting on the side panels of the saddle (see Figure 6-18). They must also be of appropriate length for the poles and shafts of larger rigs or there will be difficulty in stopping the vehicle. The best length will vary from vehicle to vehicle, and the best way to gauge it is to consult someone with experience until you learn how to judge for yourself.

Neck collars should only be used with vehicles equipped with whiffle trees that move with the movement of the shoulders. Otherwise painful sores may develop on the horse's shoulders, causing unsafe behavior in even the most well meaning of animals. The collar should be fitted as high on the shoulders as possible without interfering with the air passages. This can be checked by being sure that you can pass the flat of your hand between the collar and the horse's windpipe at the bottom of the collar

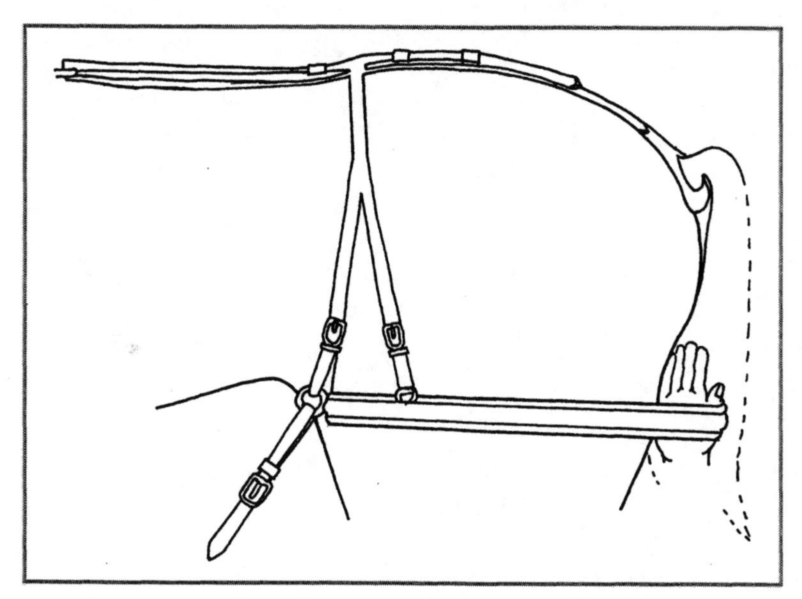

Figure 6-16 The width of the hand under the breeching strap

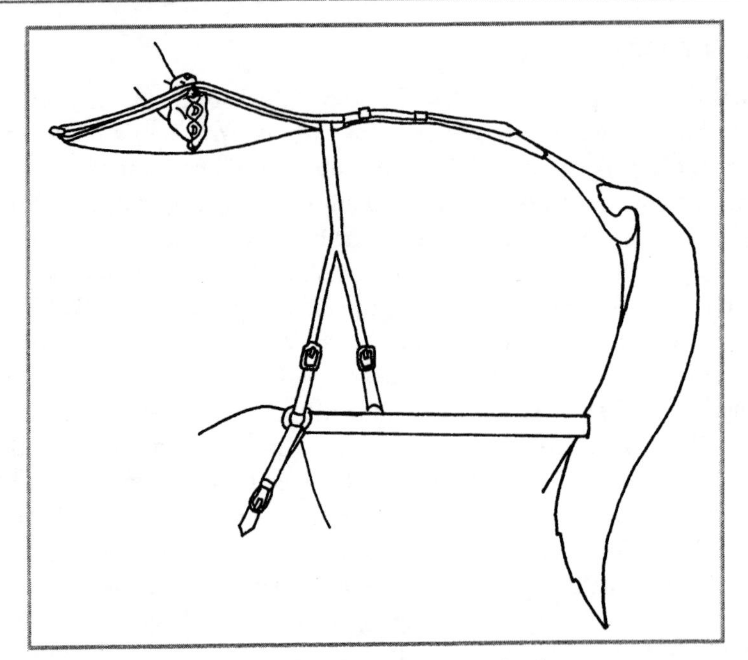

Figure 6-17 The hand under the backstrap

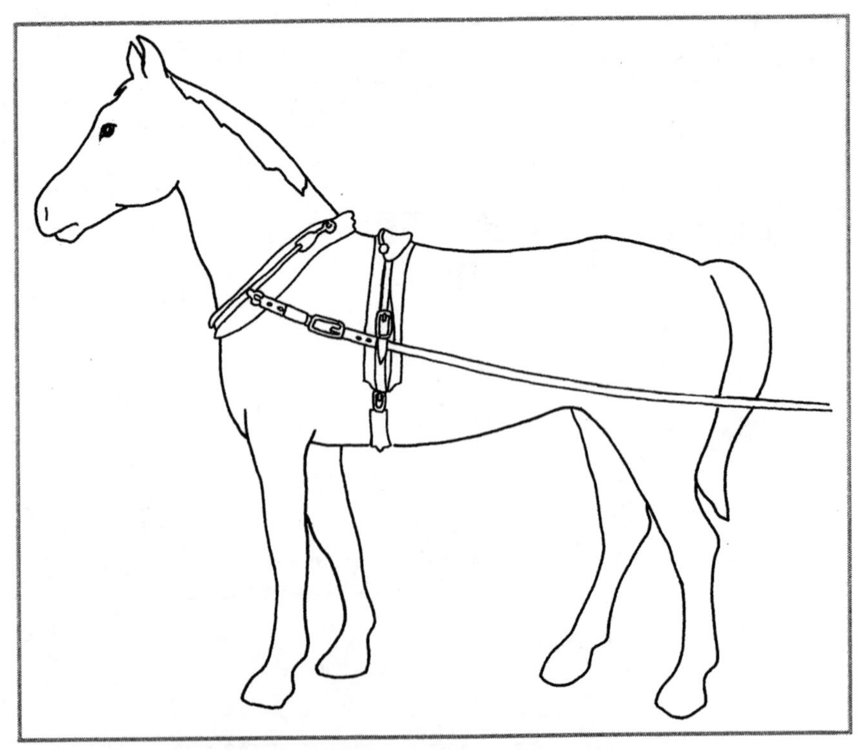

Figure 6-18 Traces properly adjusted

(see Figure 6-19). The collar should also be wide enough to spread the pressure over a suitable area when pulling. When using breast collars, adjust the neck straps and martingales (if any) so that the breast collar lies above the point of the shoulder yet below the windpipe (see Figure 6-20).

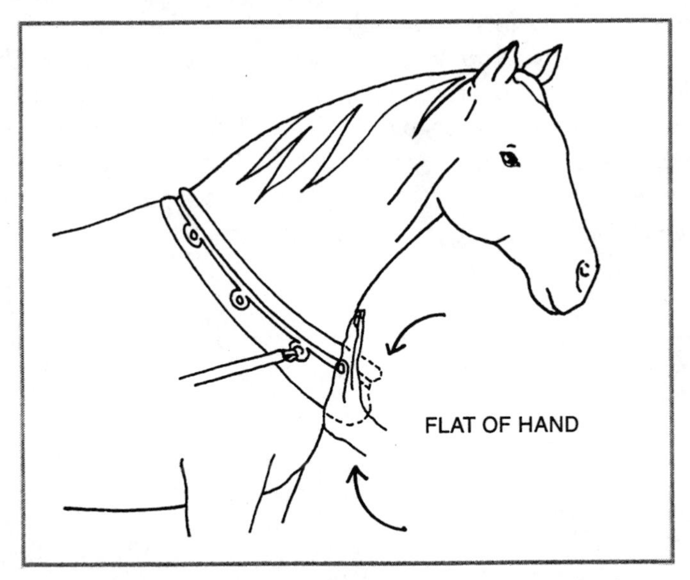

Figure 6-19 Make sure that the collar does not interfere with the windpipe.

Figure 6-20 Proper positioning of the breast collar

SUMMARY

It is beyond the scope of this book to list all possible tack items. The many uses of horses have created so many different forms of tack that such a task would be formidable. Fortunately there are some common denominators upon which we can focus. Tack usually causes unsafe conditions when it either loses physical integrity or fits so poorly that it creates pain. With the increase in the number of materials used in the production of tack, it is becoming more difficult to judge when tack will fail physically. It is now up to the user to be knowledgeable about the material used in his or her tack and how to estimate when it will fail. How to judge the fit of tack is much the same regardless of materials. It is often best to seek the guidance of experienced people until you have had enough time to gain the required knowledge.

CHAPTER

Trailering

Transporting horses is an interesting activity when things go well—and quite exciting when things do not. It can be done prudently and well, reducing risks to a minimum, or it can be done foolishly, increasing the chance of injury to everyone involved. It pays for you to develop a prudent plan or system for moving your horses, particularly if you have to transport them frequently. Said plan should include as many safeguards for human workers as possible. Trouble for humans usually comes from one or more of four sources: the training of the horse (or lack thereof); loading procedures; the physical condition of the trailer and towing vehicle; and the skill of the driver.

Training of the Horse

The first step to safety is to have a horse that enjoys trailers and is willing to load in a calm fashion. Many excellent trainers completely overlook trailer training, however. I'm not sure of the reason for this, but it certainly occurs more frequently than it should. Consider trailers from the horse's perspective for a moment. Thousands of years of instinct tell the horse that small, dark places are to be avoided. After all, a horse's best defensive strategy is to outrun a threat, and it takes room to do that. Small dark places (like caves, etc.) are often the dwelling places of predators. So there are no good reasons for a horse to walk into a confined dark space such as a trailer. It is therefore up to us to teach the horse that there is nothing to fear from being in a trailer and that good things will happen when it is in one.

Horses are often loaded using harsh or compulsive methods. Although this gets the animal into the trailer, it does not teach the horse to enjoy it. One method to do so involves feeding the horse every meal in or as near to a trailer as possible until the horse teaches itself to load in order to eat. This is just one example, but a complete discussion of training techniques is beyond the scope of this book. However, these methods are not difficult to discover. Most good trainers have a favorite they are willing to share, and nearly all of them work. The point is to find one and use it long before you have to transport your horse. Then you have an animal that is calm and willing to load when the time comes. Nothing keeps you safer than that.

Loading Procedures

The Cooperative Horse

As stated earlier, life is much safer with a calm, cooperative horse. Select the safest possible location for loading, preferably away from traffic and other distractions. Check the trailer to be sure that the wheels are securely blocked, that the hitch is either connected to a towing vehicle or weighted properly, and that the ramp (if present) is solidly positioned to eliminate any movement when the horse steps on it. Many horses become frightened if the trailer moves in any manner as they step up into it. Not only can that create a safety hazard if the horse presses the handler against the wall or steps on him or her, but it makes it more difficult to get the horse back into the trailer next time. Make the interior of the trailer as bright as possible by opening any doors or windows in the front of it. Also be sure that all doors are secured so that they will not accidentally swing and make noise or strike the horse while loading is taking place. Check all protective leg wraps and other equipment on the horse to see that it will not fail during loading and have the lead shank in the proper configuration for the particular horse you are loading (see Chapter 2 for a discussion of lead shanks and restraints). Have an escape plan in mind so that you can react faster if something goes wrong. Know where the nearest safe place or exit is at all times. It might be an open door in trailers that have one in the front, or it may be the space on the other side of the partition down the center of the trailer, or just the space in front of the chest bars in the front of the trailer. With a good horse, you probably will not need such a plan, but it is good to have one all the same. Even a good horse can jump or panic if frightened or stung by a bee.

After you have made these preparations, lead the horse confidently up and into the trailer without looking back at it. Looking back tends to slow horses down and brings some to a complete stop. This would not be true if you have specifically trained the horse to move forward while you look back at it, of course, as some trainers do. You can vary the speed to suit the horse, but the confidence with which you lead is critical. Don't expect a horse to follow a leader who looks insecure. After the horse is in, step to a safe area and secure the lead to the appropriate tieing point, using a quick release knot (see Chapter 3 for a discussion of tieing). As soon as the horse's rump passes the butt chain or butt bar in the back of the trailer, your helper (should you be lucky enough to have one) should raise and secure the chain or bar.

If there is no escape door in front or other safe area for you to use, the best approach is to lead the horse to the trailer but not to enter it yourself. You stop just short of entering and, releasing the lead, allow the horse to continue in without you. Someone else must catch the horse on the other end or you must get the butt chain or bar up to keep the horse from coming back out of the trailer. This may require more training with some horses, but it keeps you from getting caught in the front of the trailer with no place to go. It is particularly useful for loading the second horse into trailers with no escape door in the front.

The Uncooperative Horse

The safest approach to a horse that is resisting your efforts to load it is to stop trying and wait until you have had time to train it properly. Unfortunately this is not always possible. If this option is denied you, there are some other options, each with its own set of risks.

Bribing the horse into the trailer with its favorite food is frequently successful. The trick to this technique is to guard against being bitten or run over and stepped on. Offering the food in a bucket or other container usually takes care of the biting problem. With regard to being run over, remember that you are positioned in front of the horse, in Zone A. Some horses walk smoothly on, but others hesitate, feeling severe conflict, walk in place, and then, having made the decision to follow, launch themselves after you. You need to be able to move out of their way. If you can, all is well. Consequently this technique should only be employed when you have some form of escape door in the front, room in front of the chest bars, space on the other side of the center partition that you can reach quickly, or some other safe area you can use in a similar fashion.

The possibility of the horse rearing is considerable with the following methods. When a horse does not want to go forward, feels that it cannot go back, and yet feels compelled to go somewhere, it is not unreasonable for it to try to go up. Consequently, in all of the following methods, the handler should be prepared to deal with rearing by having a lead long enough to allow distance between horse and human, and a safe area to step to so that the front end of the horse does not strike or come down on top of him or her.

Use of a rump rope (see Figure 7-1) can help many horses, particularly young ones. The major precautions to take are those for not getting stepped on or run over if the horse hesitates and then decides to follow. (These are listed earlier in the section on bribing.) In lieu of a rump rope, some people like to lock arms behind the horse's rump and push forward to assist the handler. Although this works well in many instances, some horses have been known to kick with the hind feet during the procedure. Because the people cannot choose their location, they should stay as close to the horse's body as possible, so as to jam the kick before it develops power. A better approach is to have each of them hold the end of a rope or lunge line so that they can move back out of kicking range.

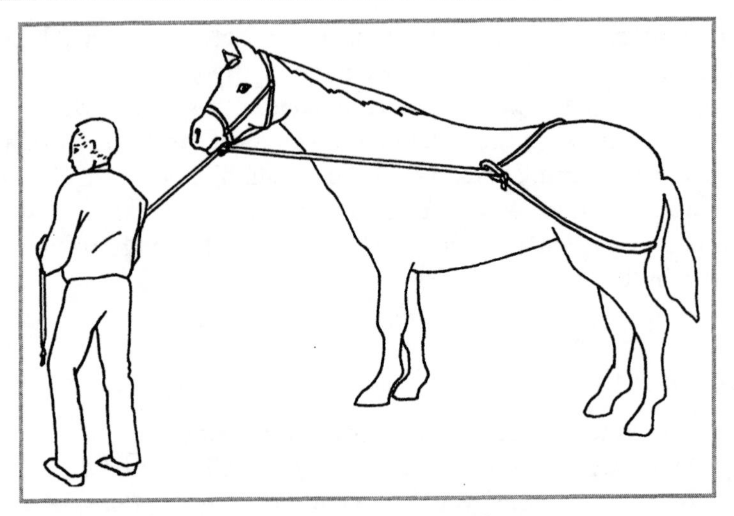

Figure 7-1 The rump rope in use

A blindfold has been beneficial with many horses. Apparently if they do not have to look at the scary object, it is easier to walk up to. If the horse stumbles momentarily on the ramp or step-up portion of the trailer it may jump in the handler's direction, so he or she again must have a safe escape area.

Some horses will walk backward onto a trailer when they will not enter it in the forward direction under any form of encouragement. When attempting to back a horse on, the handler should be extremely careful to steer the horse well. Backing the horse off the side of the ramp or into the center partition or side door can result in the horse jumping in the direction of the handler, so he or she should have a safe area in mind to jump to quickly. Fortunately the handler usually has plenty of room to maneuver when backing the horse on.

Tickling the horse's rear legs with a long-handled whisk broom (ceasing the tickling when the horse moves forward and reapplying it when it stands still or moves backward) is also a popular technique (see Figure 7-2). With this procedure, the handler must guard against being stepped on or run over, as described earlier, and the broom handler must be certain that the broom handle is long enough to keep him or her out of kicking range. Being tickled consistently can be quite annoying to some horses, and they will strike out with the hind feet, much as they would if there were flies on them. Some people swing the broom like a baseball or cricket bat, which does not damage the horse but is more stimulating than tickling. In this case, greater caution should be used with regard to the hind feet kicking out.

Some people teach the horse to walk forward even when they are looking back at it by tapping the top of the horse's rump with a long stick or dressage crop and stopping the tapping when the horse makes some sort of forward movement. They then stand by the entrance and let the horse walk past them onto the trailer (see Figure 7-3). When using this method, the main precaution to take is to keep a good hold of the horse's

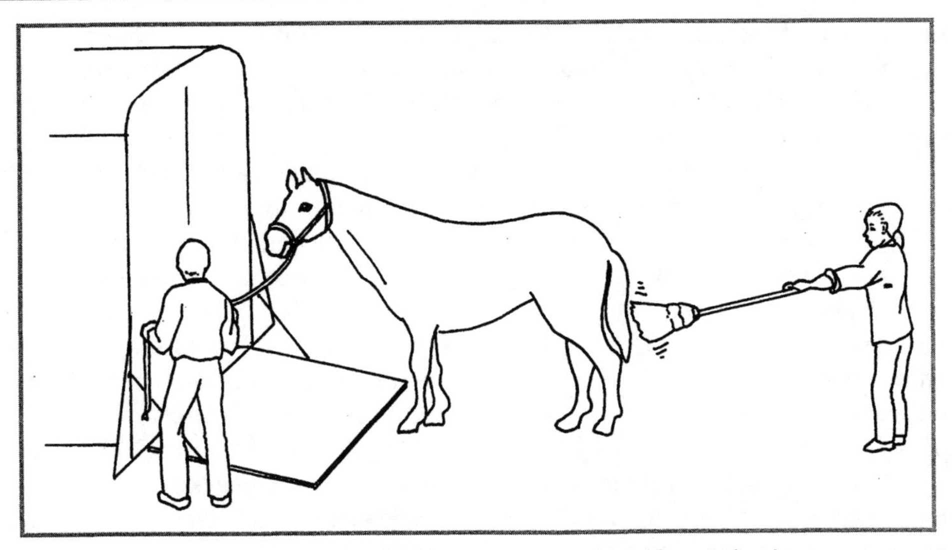

Figure 7-2 Staying back clear of the legs while tickling with a broom

Figure 7-3 The horse walking past the handler on to the trailer

head until it is loading itself. If the horse breaks free, it could turn and kick with the hind end to protest the tapping. If the head is controlled properly, it is difficult for the horse to do this.

Tranquilizers have been used successfully in many instances. You should always consult your veterinarian before using such chemicals, however, and they do not release you from safety precautions. Sedatives must be administered according to the horse's weight, and sometimes people misjudge. Some horses have negative reactions to sedatives, and sometimes the sedative wears off sooner than expected. Consequently, the handler should observe all reasonable safety precautions even when he or she is sure that it's not necessary.

Many people like to use ropes or lines to winch reluctant horses into trailers. Others use them as a gentle restraint to keep the horse from moving away from the trailer. The merits and demerits of using ropes have been argued extensively, and it is not likely that those who work with horses will agree on how or even if they should be used, at least not in the near future. If and when they are used, however, some precautions are necessary. They should not be used on certain horses. For some reason, certain horses fight you to the death when ropes are used, but be quite tractable when you use other options. For this reason, it is safest to use ropes and lines as a last resort, and only when all else has failed. Humans should always wear gloves when using lines or ropes with horses because the chances of rope burns or damage to the flesh on the hands increases in these situations. Anyone positioned in front of the horse should have a safe escape area. Care should be taken not to get tangled up in the ropes or lines if horses spin, rear, or throw themselves over on their sides to evade them. When confronted by the inevitability of entering the trailer, some frightened horses rear up so high that they fall over backward, so people standing behind the horse must guard against that as well as being kicked.

On rare occasions, you may have to load an unreasonably difficult horse without the aid of sedatives. For example, I once had a four-year-old horse sent in to training class that had never been touched by humans unless it was roped and tranquilized, mostly for a quick vet check and trimming of the hooves. I do not use sedatives in class, and it took me three days just to desensitize the horse to the point where I could place a hand on it quietly. To attempt to actually "handle" the horse during this period would produce severely fearful reactions that were quite dangerous for the humans involved. It was not the horse's fault, but it wasn't ours either, and the question is how would you load a horse like this if it were allergic to sedatives? I became quite curious as to how this horse got to campus in the first place. It turns out that the transporter, not willing to risk tranquilizers, yet recognizing the risk such a horse offers to humans, elected not to touch it at all. The trailer was backed up to the gate of the pasture in such a way that when the door was opened it formed a chute leading into the trailer. The horse was then gently herded onto the trailer loose and the doors closed. The driver went very slowly and carefully all the way with the horse loose in the trailer. When they reached campus, they found an empty turnout paddock, backed the trailer up to the gate, and gently herded the horse off the trailer into the paddock. The entire trip was made without a human coming close enough to the horse to be kicked. Please understand that this is not my favorite method of trailering. It has its faults, but if you are faced with a situation where you must transport, yet the chances of being injured are unacceptably high, it would be better to do this than to sustain injuries.

Unloading Procedures

The Cooperative Horse

Again, choose the safest possible location for unloading, away from traffic and other distractions. Be certain that the trailer is properly blocked and that the ramp (if present) is properly based on the ground so that it doesn't wobble when the horse steps on it. Secure any open doors so that they won't swing and bang into the horse during unloading. Clear the area behind the trailer of any innocent bystanders, because some horses come off the trailer quickly if they become frightened. Try to release the head before lowering the butt chain or bar. If the butt chain or bar is taken down with the head still tied, there is a chance the horse may anticipate your intentions and move backward sooner than you would like. If it is still tied in front, it may panic when it feels the unexpected pressure on its head from the halter. It may then throw itself backward trying to escape the pressure, which of course only creates more. This downward spiral usually creates unsafe conditions and can be avoided by releasing the head first.

As the horse moves backward off the trailer, steer as well as possible. If the horse bumps into the center partition, it may jump back in the direction of the handler, so a safe area will again be needed. If the horse goes off the side of the ramp, it may jump back toward the handler or it may continue quickly off the trailer, possibly rearing before it does so. Should rearing occur, the handler must be ready to move back out of striking distance.

The Uncooperative Horse

Unloading the difficult horse is often a question of managing its speed. Horses that don't want to be on the trailer in the first place are often more than willing to come off, they're just in too much of a hurry. Some seem to panic, throw their heads up, hitting the roof of the trailer, and fly off backward, sometimes falling down and thrashing around as they do so.

The main precautions are that of having a clear area behind the trailer and not getting in the horse's way as it comes off. A long lead shank is nice if the horse falls or decides to rear after exiting; a protective helmet for the horse's head minimizes the fear created when it hits its head on the roof. Should the horse fall, stay clear of the flailing legs until the horse rights itself.

Some horses that would otherwise panic remain calm if they are allowed to turn around and exit walking forward. This requires a trailer wide enough for the turn and sometimes that the other horses be unloaded first to provide the necessary space. If your horse is uncomfortable coming off the trailer, this is an option worth considering.

The Bottom Line for Loading

Though it is worth learning all the different methods for loading difficult horses, one point should be reemphasized. The safest way to load horses is to have calm, well-trained horses that are comfortable in trailers and simply walk on every time you ask them to. This behavior should be part of every prepurchase exam and fostered by everyone who raises horses. Certainly there will be exceptions due to trailer accidents and other unusual events, but the fact remains that there are many horses that load well, and if safety is your top priority, you should settle for nothing less.

Physical Condition of the Vehicles

The Towing Vehicle

Towing vehicles come in many forms and sizes. They should be large and powerful enough to handle the weight of the trailer should you have to pull uphill or go slowly downhill for long periods of time. Many a transmission has been ruined and many a motor has been burned out over the years by hitching trailers to towing vehicles that were not powerful enough for the job. In some instances this has caused accidents that have injured people. Naturally the towing vehicle needs to be in good repair, and the steering, brakes, and lights need to be checked frequently. The hitching ball and its supports should be solid enough to handle the weight of the trailer and of the proper height to match it. Most people prefer to have the hitch at a height that keeps the front section of the trailer parallel to the ground.

The Trailer

The trailer itself must also be in good repair. The hitch should function well and be equipped with a locking pin to prevent the hitch from popping open, and a safety chain in case the hitch fails. Goose neck or fifth wheel attachments may not need chains but should be checked to ensure that they are in good shape. It is an advantage to have brakes installed in the trailer that automatically activate just prior to those in the towing vehicle when the brake pedal in the towing vehicle is depressed. This slows the trailer slightly before the towing vehicle and helps to prevent jackknifing, which can be quite hazardous. However, trailer brakes can malfunction like anything else and should be checked frequently. Many people with standard hitches prefer to have sway bars attached from the towing vehicle to the trailer to minimize sideways movement between the two vehicles. The inside of the trailer should be checked for sharp edges, and the floor boards should be inspected for functional quality. The brake and signal lights of the trailer should be checked before each trip.

SKILL OF THE DRIVER

Trailer rigs should not be driven like ordinary vehicles. To be successful, a driver must think differently when transporting horses. He or she must stop and start gradually and avoid abrupt changes in speed because these are magnified back in the trailer and cause bad behavior on the part of the horse, which in turn creates risk for the handlers when it is time to unload. Curves in the road should be negotiated slowly and smoothly for the same reason. Bumps in the road should be taken at slower speeds than normal because they are worse in the trailer than they are in the comfortable towing vehicle. Following distances need to be greater when towing horses because the rig is heavier than a single vehicle and needs more room to stop. Greater following distance is also the best way to avoid sudden changes in speed when braking, which in turn is the best way to avoid jackknifing the vehicles and causing a severe accident.

Drivers occasionally create hazards doing things they normally would not do because they do not want to have to back up the trailer to turn around, or back the trailer into a small parking space. Backing up is an important skill for drivers and should be mastered by anyone who intends to ship horses frequently. To avoid dangerous situations, this is best learned before trips, not during them.

SUMMARY

Driving on the open road is unpredictable enough even without a trailer. This unpredictable environment creates enough risks without you looking for new ones. It therefore pays for you to be careful with regard to known sources of trouble. This involves (but is not limited to) having a calm horse that is well trained to load, knowledge of good loading techniques, physically sound vehicles, and a good driver. Nothing can guarantee success, but with these items you are well prepared to face the uncertainties of shipping.

Appendix

Table A-1 Breed Registry Associations

Akhal Teke Registry of America
Rt. 5, Box 110
Staunton, VA 22401-8906

American Andulasian Horse Assn.
6990 Manning Road
Economy, IN 47339-9736

American Baskir Curly Registry
P.O. Box 246
Ely, NV 89301-0246

American Buckskin Registry Assn.
P.O. Box 3850
Redding, CA 96049-3850

American Connemara Pony Society
2630 Hunting Ridge
Winchester, VA 22603

American Cream Draft Horse Assn.
P.O. Box 2065 Noble Avenue
Charles City, IA 50616-9108

American Dartmoor Pony Assn.
15870 Paseo Mantra Road
Anna, OH 45302

American Dominant Gray Registry
10980 "8" Mile Road
Battle Creek, MI 49017-9560

American Exmoor Pony Registry
c/o American Livestock Breeds
Conservancy
P.O. Box 477
Pittsboro, NC 27312-0477

American Hackney Horse Society
#A 4059 Iron Works Road
Lexington, KY 40511-8462

American Hanoverian Society
4059 Iron Works Pike
Lexington, KY 40511

American Holsteiner Horse Assn.
#1 222 East Main Street
Georgetown, KY 40324-1712

American Horizon Horse Registry
P.O. Box 564
Belen, NM 87002-0564

American Indian Horse Registry
Rt. 3, Box 64
Lockart, TX 78644

American Miniature Horse Assn.
2908 SE Loop 820
Fort Worth, TX 76140-1073

American Miniature Horse Registry
P.O. Box 3415
Peoria, IL 61614-3415

American Morgan Horse Assn.
P.O. Box 960
Shelburne, VT 05482-0960

American Mustang and Burro Assn.
P.O. Box 788
Lincoln, CA 95648

Table A-1 Breed Registry Associations *(continued)*

American Mustang Assn.
P.O. Box 338
Yucaipa, CA 92399

American Paint Horse Assn.
P.O. Box 961023
Fort Worth, TX 76161-0023

American Quarter Horse Assn.
P.O. Box 200
Amarillo, TX 79168-0001

American Quarter Pony Assn.
P.O. Box 30
New Sharon, IA 50207

American Saddlebred Horse Assn.
4093 Iron Works Pike
Lexington, KY 40511-8434

American Shetland Pony Club
P.O. Box 3415
Peoria, IL 61614-3415

American Shire Horse Assn.
2354 315 Court
Adel, IA 50003

American Suffolk Horse Assn.
4240 Goehring Road
Ledbetter, TX 78946-9707

American Tarpan Studbook Assn.
1658 Coleman Avenue
Macon, GA 31201-6602

American Trakehner Assn.
1520 West Church Street
Newark, OH 43055

American Walking Pony Registry
P.O. Box 5282
Macon, GA 31208-5282

American Warmblood Registry (also American Warmblood Society)
6801 West Romley Avenue
Phoenix, AZ 85043

American Welara Pony Society
P.O. Box 401
Yucca Valley, CA 92286-0401

Appaloosa Horse Club
P.O. Box 8403
Moscow, ID 83843-0903

Arabian Horse Registry of America
12000 Zuni Street
Westminster, CO 80234-2300

Belgian Draft Horse Corporation of America
P.O. Box 335
Wabash, IN 46992-0335

Caspian Horse Society of America
Rt. 7, Box 7504
Brenham, TX 77833

Chilean Corralero Registry International
230 East North Avenue
Antigo, WI 54409

Cleveland Bay Horse Society of North America
P.O. Box 221
South Windham, CT 06266

Clydesdale Breeders of the U.S.A.
17378 Kelley Road
Pecatonia, IL 61063

Falabella Miniature Horse Assn. of America
P.O. Box 3036
125 Glenwood Drive
Gettysburg, PA 17325

Table A-1 Breed Registry Associations *(continued)*

Florida Cracker Horse Assn.
P.O. Box 186
Newberry, FL 32669-0186

Friesian Horse Assn. of North America
4127 Kentridge Drive SE
Grand Rapids, MI 49508-3705

Galiceno Horse Breeders Assn.
Box 219
Godley, TX 76044-0219

Golden American Saddlebred Horse Assn.
4237 30th Avenue
Oxford Junction, IA 52323-9724

Haflinger Assn. of America
14570 Gratiot Road
Hemlock, MI 48626-9416

Half Quarter Horse Registry of America
29264 Bouquet Canyon Road
Sangus, CA 91350

Half Saddlebred Registry
319 South Sixth Street
Coshocton, OH 43812-2119

International Arabian Horse Assn.
(also includes Half-Arab and Anglo-Arabian registries)
P.O. Box 33696
Denver, CO 80233-0696

International Arabian Horse Registry of North America
P.O. Box 325
Delphi Falls, NY 13501-0325

International Buckskin Horse Assn.
P.O. Box 268
Shelby, IN 46377-0268

International Colored Appaloosa Assn.
P.O. Box 4424
Springfield, MO 65808-4424

International Morab Breeders Assn.
(up to 75% Arabian or Morgan)
South 101 West 34628 Highway 99
Eagle, WI 53119-1857

International Plantation Walking Horse Assn.
P.O. Box 510
Haymarket, VA 22069-0510

International Sporthorse Registry and Oldenburg Verband N.A.
P.O. Box 849
Streamwood, IL 60107

International Trotting and Pacing Assn.
575 Broadway
Hanover, PA 17331-2007

Jockey Club, The
821 Corporate Drive
Lexington, KY 40503-2794

Lippizan Assn. of North America
P.O. Box 1133
Anderson, IN 46015-1133

Missouri Fox Trotting Horse Breed Assn.
P.O. Box 1027
Ava, MO 65608-1027

Mountain Pleasure Horse Assn.
P.O. Box 670
Paris, KY 40362-0670

National Pinto Arabian Registry
942 Kathryn Lane
Royse, TX 75189

Table A-1 Breed Registry Associations *(continued)*

National Pinto Horse Registry
P.O. Box 486
Oxford, NY 12820-0486

National Spotted Saddle Horse Assn.
P.O. Box 898
Murfreesboro, TN 37133-0898

New Forest Pony Assn.
P.O. Box 206
Pascoag, RI 02859

North American District of the Belgian Warmblood Breeding Assn.
General Hunton Road
Broad Run, VA 22014-4877

North American Draft Cross Assn.
742 Rebecca Avenue
Westerville, OH 43081

North American Exmoors
RR 4 Box 273
Amherst, Nova Scotia, B4H 3Y2
CANADA

North American Morab Horse Assn.
W. 3174 Faro Springs Road
Hilbert, WI 54129

North American Mustang Assn. and Registry
P.O. Box 850906
Mesquite, TX 75185-0906

North American Selle Francais Horse Assn.
P.O. Box 646
Winchester, VA 22604-0646

North America Shagya (Arabian) Society
2520 60th Ave., SW
Rochester, MN 55902

North American Single-Footing Horse Assn.
P.O. Box 1079
Three Forks, MT 59752-1079

North American Trakehner Assn.
1660 Collier Road
Akron, OH 44320

Norwegian Fjord Assn. of North America
24570 W. Chardon Road
Grayslake, IL 60030

Palomino Horse Assn.
HC63, Box 24
Dornsife, PA 17623

Palomino Horse Breeders of America
15253 E. Skelly Drive
Tulsa, OK 74116-2637

Palomino Ponies of America
160 Warbasse Junction Road
Lafayette, NJ 07848-9408

Paso Fino Horse Assn.
P.O. Box 600
Bowling Green, FL 33834-0600

Percheron Horse Assn. of America
P.O. Box 141
Fredericktown, OH 43019-0141

Performance Horse Registry
P.O. Box 24710
Lexington, KY 40524-4710

Peruvian Part Blood Registry
2027 Cribbens Street
Boise, ID 83704

Peruvian Paso Horse Registry of North America (and Peruvian Paso Part-Blood Registry)
#4 1038 4th Street
Santa Rosa, CA 95404-4319

Table A-1 Breed Registry Associations *(continued)*

Pintabian Horse Registry
P.O. Box A
Karlstad, MN 56732

Pinto Horse Assn. of America
1900 Samuels Avenue
Fort Worth, TX 76102-1141

Pony of the Americas Club
5240 Elmwood Avenue
Indianapolis, IN 46203-5990

Quarter Sport Horse Registry
1463 Country Lane
Bellingham, WA 98225-8515

Racking Horse Breeders Assn. of America
Rt. 2, Box 72-A
Decatur, AL 35603

Ridden Standardbred Assn.
1578 Fleet Road
Troy, OH 49373

Rocky Mountain Horse Assn.
1140 McCalls Mill Road
Lexington, KY 40515

Royal Warmblood Studbook of the Netherlands
North American Department
P.O. Box 828
Winchester, OR 97495-0828

Spanish-Barb Breeders Assn. International
12284 Springridge Road
Terry, MS 39170

Spanish Mustang Registry
Rt. 3, Box 7670
Wilcox, AZ 85643

Spanish-Norman Registry, Inc.
P.O. Box 985
Woodbury, CT 06798

Standardbred Pleasure Horse Organization
31930 Lambson Forest Road
Galena, MD 21653

Swedish Gotland Breeders' Society
Rt. 3, Box 134
Corinth, KY 41010-9010

Swedish Warmblood Assn. of North America
P.O. Box 1587
Coupeville, WA 98239-1587

Tennessee Walking Horse Breeders' and Exhibitors' Assn.
P.O. Box 286
Lewisburg, TN 37091-0286

Thoroughbred Horses for Sport
P.O. Box 160
Great Falls, VA 22066

Thoroughbred in Sport Assn.
964 Gale Drive
Wisconsin Dells, WI 53965

United Quarab Registry
31100 NE Fernwood Road
Newbury, OR 97132-7012

United States Icelandic Horse Federation
38 Park Street
Montclair, NY 07042

United States Lippizan Registry
13351 DP Chula Road
Amelia, VA 23002

Table A-1 Breed Registry Associations *(concluded)*

United States Trotting Assn.
750 Michigan Avenue
Columbus, OH 43215-1191

Universal Perkehner Society
P.O. Box 1874
Cave Creek, AZ 85311-1874

Walkaloosa Horse Assn.
3815 North Campbell Road
Otis Orchards, WA 99027

Walking Horse Owners' Assn. of America
#3A 1535 West Northfield Blvd.
Murfreesboro, TN 37129

Welsh Pony and Cob Society of America
P.O. Box 2977
Winchester, VA 22604-2977

Westfalen Warmblood Assn. of America
18432 Biladeau Lane
Penn Valley, CA 95946

Table A-2 National and International Horse Organizations

American Driving Society
P.O. Box 160
Metamora, MI 48455-0160

American Granprix Assn.
3104 Cherry Palm Drive
Suite 220
Tampa, FL 33619

American Horse Shows Assn., Inc.
220 E. 42nd Street, #409
New York, NY 10017-5876

American Hunter and Jumper Foundation
340 E. Hillendake Road
Kennett Square, PA 19348

American Paint Horse Assn.
P.O. Box 961023
Fort Worth, TX 76161-0023

American Polocrosse Assn.
250 Morning Glory Lane
Durango, CO 81301

American Quarter Horse Assn.
P.O. Box 200
Amarillo, TX 79168-0001

American Royal Assn.
1701 American Royal Court
Kansas City, MO 64102

American Saddlebred Grand National
4093 Iron Works Pike
Lexington, KY 40511-8434

American Team Penning Assn.
1776 Montano Road, NW, Building #3
Albuquerque, NM 87107

American Vaulting Assn.
642 Alford Place
Bainbridge Island, WA 98110-4608

American Warmblood Society
6801 W. Romley Avenue
Phoenix, AZ 85043

Appaloosa Horse Club, Inc.
P.O. Box 8403
Moscow, ID 83843-0903

Appaloosa Horse Club of Canada
Box 940
Claresholm, AB T0L 0T0
CANADA

Arabian Breeders' Marketing Network Augusta Futurity
Atlantic Coast Cutting Horse Association
P.O. Box 936
Augusta, GA 30903-0936

Azteca Association of Canada
R.R. 2
Paris, ON N3L 3E2
CANADA

Barrel Futurities of America, Inc.
4701 Parsons Road
Springdale, AR 72764

Canadian Belgian Horse Assn.
R.R. 3
Schomberg, ON L0G 1T0
CANADA

Canadian Buckskin Assn.
P.O. Box 135
Okotoks, AB T0L 1T0
CANADA

Canadian Cutting Horse Assn.
234 17th Ave., NE
Calgary, Alberta T2E 1L8
CANADA

Table A-2 National and International Horse Organizations *(continued)*

Canadian Dressage Owners & Riders Association
R.R. 2
Millbrook, ON L0A 1G0
CANADA

Canadian Driving Society
40774 Taylor Road
De Roche BC V0M 1G0
CANADA

Canadian Fjord Horse Assn.
Box 411
Dauphin, MB R7N 2V2
CANADA

Canadian Hackney Society
R.R. 1
Linsay, ON K9V 4R1
CANADA

Canadian Haflinger Assn.
General Delivery
Shagulandah, ON P0P 1W0
CANADA

Canadian Horse, The
Upper Canada District
1289 Pilon Road
Clarence Creek, ON K0A 1N0
CANADA

Canadian Icelandic Horse Federation
R.R. 1
Vernon, BC V1T 6L4
CANADA

Canadian Percheron Assn.
Box 200
Crossfield, AB T0M 0S0
CANADA

Canadian Shire Horse Assn.
1882 Conc. Road 10
Blackstock, ON L0B 1B0
CANADA

Carriage Assn. of America, Inc., The
RD 1, Box 115
Salem, NJ 08079

Clydesdale Horse Assn. of Canada
R.R. 2
Thomton, ON L0L 2N0
CANADA

Del Mar National Horse Show
2260 Jimmy Durante Blvd.
Del Mar, CA 92014

Gladstone Equestrian Assn.
P.O. Box 119
Gladstone, NJ 07934

Golden American Saddlebred Horse Assn.
4237 30th Ave.
Oxford Junction, IA 52323-9724

Hanoverian Horse Society
R.R. 2
Elora, ON N0B 1S0
CANADA

Intercollegiate Horse Show Assn.
P.O. Box 741
Stoney Brook, NY 11790-0741

International Arabian Horse Assn.
P.O. Box 33696
Denver, CO 80233-0696

International Buckskin Horse Assn.
P.O. Box 268
Shelby, IN 46377-0268

International Halter-Pleasure Horse Assn.
256 N. Highway 377
Pilot Point, TX 76258-9624

Table A-2 National and International Horse Organizations *(continued)*

International Hunter Futurity
P.O. Box 13244
Lexington, KY 40583-3244

International Jumper Futurity
P.O. Box 2830
Roseville, CA, 95746-2830

International Side-Saddle Org., The
P.O. Box 282
Albany Bay, NH 03810-0282

Japan Racing Assn.
New York Office
399 Park Avenue
27th Floor
New York, NY 10022

Malayan Racing Assn.
Paddock Block
Bukit Timah Racecourse
Singapore 1128

Masters of Foxhounds Assn. of America
Route 3, Box 51
Morven Park
Leesburg, VA 22075

National Barrel Horse Assn.
725 Broad Street
Augusta, GA 30901-1305

National Cutting Horse Association
4704 Hwy. 377 S.
Fort Worth, TX 76116-8805

National Grand Prix League
2508 Keller Pkwy.
St. Paul, MN 55109

National Horse Show Commission
Route 1, Box 257
Graham, AL 36263-9519

National Hunter and Jumper Assn., The
P.O. Box 1015
Riverside, CT 06878-1015

National Reining Horse Assn.
448 Main Street, #204
Coshocton, OH 43812-1200

National Snaffle Bit Assn.
1 Indiana Square, #2540
Indianapolis, IN 46204

North American Riding for the
Handicapped Assn.
P.O. Box 33150
Denver, CO 80233

Palomino Horse Assn.
Box 24, Star Route
Dornsife, PA 17823

Palomino Horse Breeders of America
15253 E. Skelly Drive
Tulsa, OK 74116-2637

Professional Horsemen's Assn. of
America
20 Blue Ridge Lane
Wilton, CT 06897-4127

Pyramid Society, The
P.O. Box 11941
Lexington, KY 40579

Ride and Tie Assn., The
1865 Indian Valley Road
Novato, CA 94947

Societe Des Eleveurs De Checaux
Canadiens
68 rue Deslauriers
Pierrefonds, PQ H8Y 2E4
CANADA

Table A-2 National and International Horse Organizations *(concluded)*

Special Olympics International
1350 New York Ave., NW, #500
Washington, DC 20005-4709

Tennessee Walking Horse National Celebration
P.O. Box 1010
Shelbyville, TN 37160-1010

United Professional Horsemens Assn.
4059 Iron Works Pike
Lexington, KY 40511-8434

United States Combined Training Assn.
P.O. Box 2247
Leesburg, VA 22075-2247

United States Dressage Federation
P.O. Box 6669
Lincoln, NE 68506-0669

United States Equestrian Team
Pottersville Road
Gladstone, NJ 07934

United States Olympic Committee
One Olympic Plaza
Colorado Springs, CO 80909

United States Polo Association
4059 Iron Works Pike
Lexington, KY 40511-8434

United States Team Penning Assn.
P.O. Box 161848
Fort Worth, TX 76161-1848

United States Team Roping Championships
P.O. Box 7651
Albuquerque, NM 87194

United States Vaulting Federation
RD 1, Box 235
Pittsown, NJ 08867-9722

Index

V

W

Y

Z